Introduction to 3D Data

Modeling with ArcGIS® 3D Analyst™ and Google Earth™

K. Heather Kennedy

JOHN WILEY & SONS, INC.

This book is printed on acid-free paper. ♾

Published by John Wiley & Sons, Inc., Hoboken, New Jersey
Published simultaneously in Canada

For general information about our other products and services, please contact our Customer Care Department within the United States at (800) 762-2974, outside the United States at (317) 572-3993 or fax (317) 572-4002.

Wiley also publishes its books in a variety of electronic formats. Some content that appears in print may not be available in electronic books. For more information about Wiley products, visit our web site at www .wiley.com.

Library of Congress Cataloging-in-Publication Data:

Kennedy, K. Heather.
Introduction to 3D data : modeling with arcGIS 3D analyst and Google earth / K. Heather Kennedy.
p. cm.
Includes index.
ISBN 978-0-470-38124-3 (cloth : alk. paper)
1. ArcGIS. 2. Geographic information systems. 3. Three-dimensional display systems. I. Title.
G70.212.K464 2009
005.74′3—dc22

2009009773

Printed in the United States of America

10 9 8 7

Contents

Contents

Preface

Introduction to 3D Data teaches GIS specialists, analysts, and technicians how to use ESRI's ArcGIS 3D Analyst to model and analyze three-dimensional geographical surfaces, create 3D data, and produce displays ranging from topographically realistic maps to 3D scenes and spherical earthlike views. The book is organized into 10 chapters, each focusing on one data type or software interface (ArcCatalog, ArcScene, ArcGlobe, or Google Earth). There are 39 step-by-step project exercises, with plain-language discussions throughout of pertinent data structures and software mechanics. My goal was to create a friendly, engaging atmosphere that strikes a balance between reference-like tutorials that just tell you what to do but not why, and academic tomes that celebrate theory without suggesting any real-world application. After going through these exercises, you will know exactly what 3D Analyst can do, and you will remember the situations in which you applied particular techniques and created particular types of data.

Some readers will recognize material from my previous book, *Data in Three Dimensions: A Guide to ArcGIS 3D Analyst* (Onword Press, 2004), which covered 3D Analyst for ArcGIS 8.*x*. *Introduction to 3D Data* is updated and expanded for ArcGIS 9.3 and covers new data formats, such as Terrains, multipatch features, and KML. Google Earth is also addressed, but 3D Analyst remains the focus since its strength is GIS data creation and analysis, while Google Earth is mostly for display.

You will need to have ArcView installed to do the exercises in this book, and you will need a concurrent 3D Analyst license. Most of the exercises can be done using version 9.1 or 9.2, but some require 9.3. You will also need to have Google Earth installed.

The sample data on the support website at www.wiley.com/college/kennedy is only for tutorial use. The data has been altered and so is not reliable for purposes other than illustrative or educational. The datasets are not to be sold, copied (except for personal use), or distributed.

Acknowledgments

For providing the datasets used in this book, I'd like to thank Kent Anness and Rusty Anderson at Kentucky WRIS GIS, Ryan Freedman at Lakes Environmental Software and WebGIS.com, Jason Stoker at EROS, the North Carolina Floodplain Mapping Program via the USGS "CLICK" website, John Kelly at the SLO Datafinder, and Carol Schuldt of the Kennedy Library at Cal Poly, San Luis Obispo. I'd also like to thank Jim Harper, Dan Magers, Nancy Cintron, and Jeri Freedman of John Wiley & Sons, Josh Lazarus, and Lynda Gregory. I'd like to thank my husband, Tom McMurdo, for his never-ending support and for doing my chores while I wrote nights and weekends. Finally, I'd like to thank my dad, Michael Kennedy, author of many GIS books, including *Introducing Geographic Information Systems with ArcGIS*, Second Edition (Wiley & Sons, 2009) for his encouragement and support, and for his infectious enthusiasm that makes GIS fun.

Introduction to 3D Data

CHAPTER **1**

Introduction to 3D Data: Modeling with ArcGIS 3D Analyst and Google Earth

Introduction to 3D Data is a self-study tutorial workbook that teaches you how to create data and maps with ESRI's 3D Analyst software, and to integrate them with Google Earth.

The datasets for all of the exercises in the book are provided online at www.wiley.com/college/kennedy. You must already have ArcGIS 3D Analyst installed to use this tutorial, as the book does not come with any trial software. Most of the 3D Analyst exercises can be done with versions 9.1 or 9.2 of ArcView, ArcEditor, or ArcInfo; some exercises require 9.3. Google Earth is free.

This book is designed for people who are already familiar with ESRI products, particularly ArcMap and ArcCatalog, but who would like to understand the ins and outs of the three-dimensional modeling environment. While you can do the exercises in any order, you should work through early chapters first, since instructions in later chapters are somewhat abbreviated.

3D Analyst is designed primarily to create surface elevation data and display it in three dimensions. It provides additional analysis functions such as viewshed, surface area, and volume calculation. Its original interface, ArcScene, presents data in three-dimensional space.

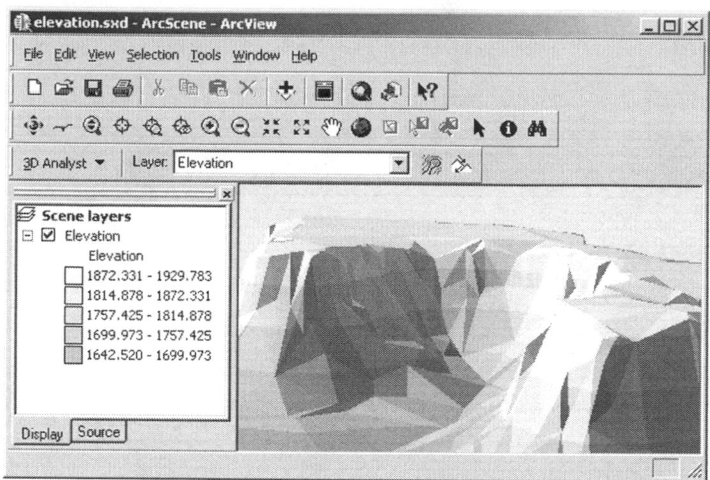

ArcScene models data in three-dimensional space

In version 9.0 ESRI added ArcGlobe to the package, which allows you to view large datasets in a global format.

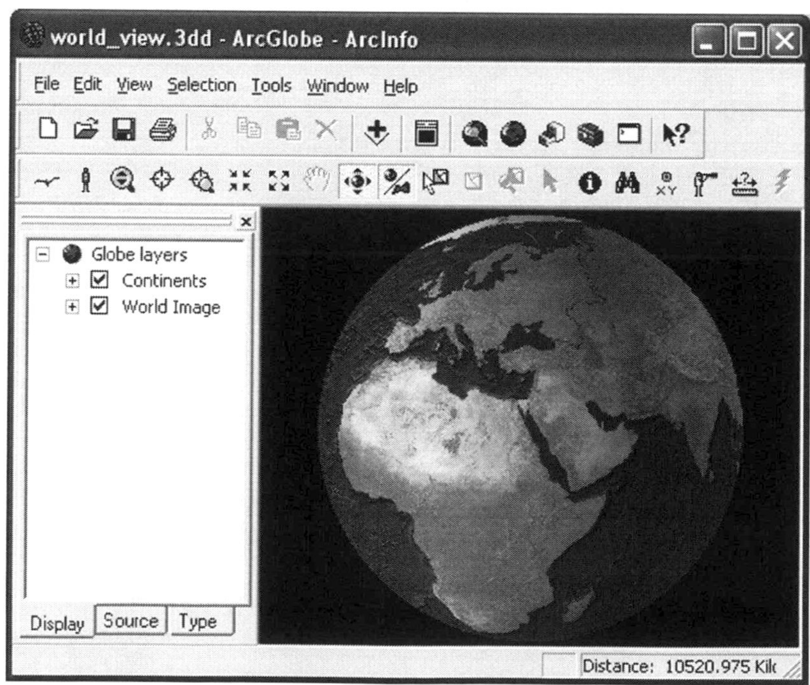

ArcGlobe models data on the earth

Increasingly, however, GIS users and the general public expect to be able to view maps interactively, on the web, for free. This is thanks largely to Google Earth, which has revolutionized the way we view spatial information. ArcGIS 3D Analyst has the power to create and analyze geographic data, but Google Earth has the speed and intuitive interface that makes it a staple for displaying maps and sharing spatial information.

With 3D Analyst, you can create TIN (Triangulated Irregular Network) and raster surface models from any vector elevation data such as contour lines, GPS points, or survey points. In ArcScene and ArcGlobe, you can drape images and vector features over surfaces, fly through your GIS data in 3D perspective, and make movies of your flights. You can extrude 2D points, lines, and polygons into lines, walls, and solids, and you can create multipatch "true 3D" features. You can calculate slope, aspect, hillshade, volume, and surface area; create contour lines, and determine visibility from any point on a surface. You can also determine lines of sight, create profile graphs of a surface, and digitize 3D features and graphics.

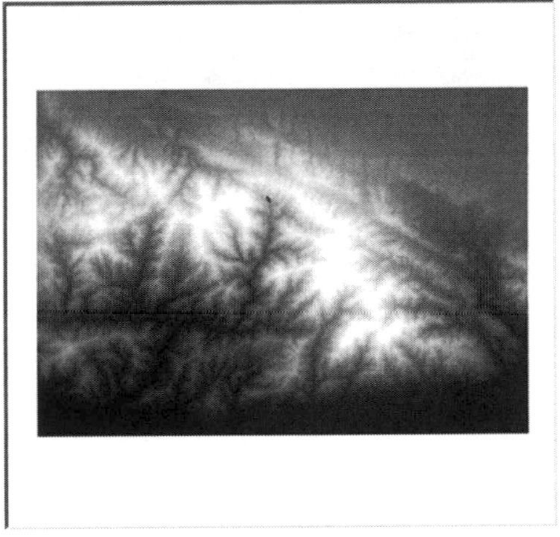

An elevation raster in ArcMap

The same elevation raster in ArcScene

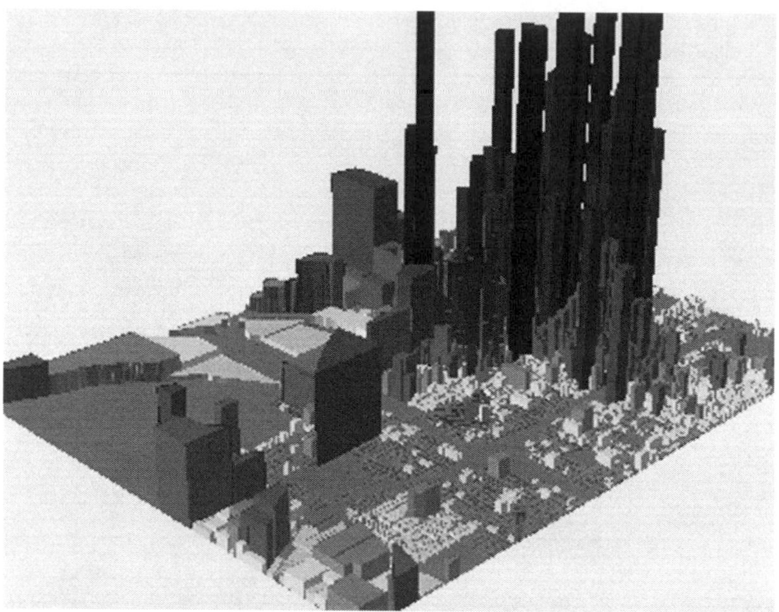

Parcels colored and extruded by land value

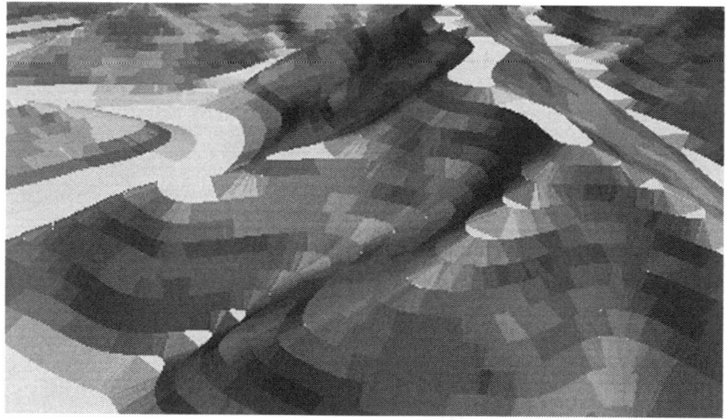

A TIN created from contour lines with faces symbolized by slope

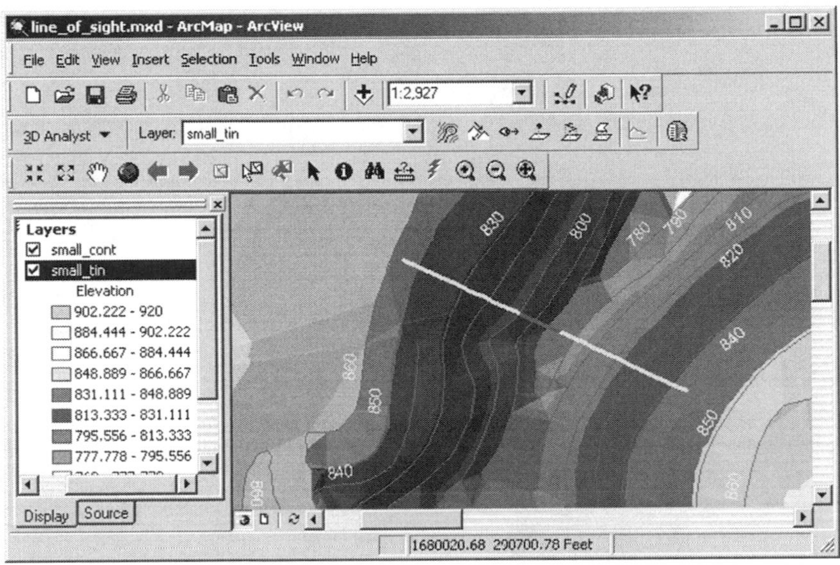

A line of sight drawn in ArcMap

3D Data Overview

X, Y, and Z Values

All geographical data contains horizontal x,y coordinate values. To work in three dimensions, you need data that contains z values as well. For each x,y location stored in a 3D dataset, a z value is stored that represents an attribute other than that location's horizontal position. In a terrain model, the z value represents elevation, or height about sea level.

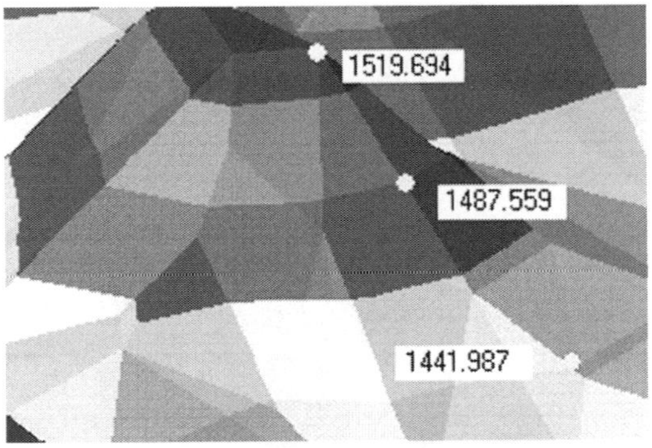

Three locations on the surface of a TIN, each labeled with their elevation (z) values in feet

3D Analyst works primarily with raster, TIN, and 3D vector feature data. Rasters and TINs are used to model surfaces, not just of terrain but of any phenomenon that varies continuously across an area, such as precipitation, chemical concentration, pollution dispersion, noise levels, population distribution, or soil pH.

Rasters

A raster represents a surface as a rectangular grid of evenly spaced square cells. Each cell is the same size and has a unique row and column address. A cell can represent a square kilometer, a square meter, or a square centimeter. The smaller the cells, the more detailed the raster, and the larger the file space taken up by the grid.

Since the grid is uniform, its horizontal (x,y) coordinates don't need to be stored in each cell. Instead they are calculated from the x,y location of the lower-left cell in the grid. Each cell does, however, hold its own z value that represents a quantity or a category of phenomena such as elevation, crop yield, or reflected light.

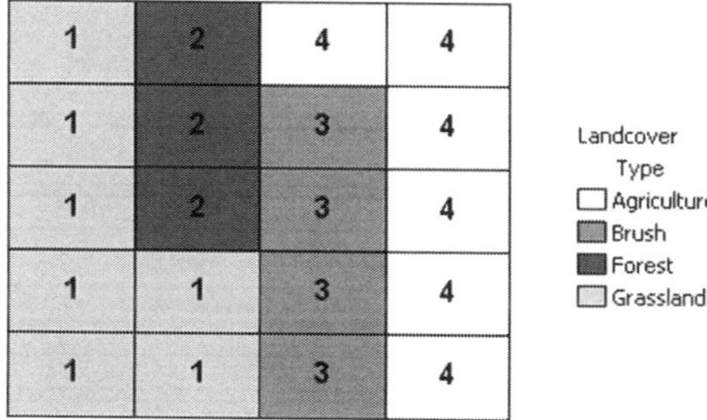

Cells in a landuse grid. All cells with the same value are symbolized by the same color

While landuse could also be represented by discrete vector polygons, vector data cannot represent values that change gradually, or continuously, over an area.

4.80	10.80	16.80	22.80
3.60	9.60	15.60	21.60
2.40	8.40	14.40	20.40
1.20	7.20	13.20	19.20
0.00	6.00	12.00	18.00

Elevation
Value
High : 1372
Low : 0.0

Cell in a continuous grid, symbolized by value range

Raster data is often divided into two categories: image and thematic. In an image, the surface phenomenon is the reflection or emission of light, or some other band in the electromagnetic spectrum, and can be measured by camera or satellite.

An aerial photograph. Cells in this raster represent light reflected from the earth's surface

When a phenomenon such as light is measured by a camera or a satellite, each cell's value represents the light and color at that point on the surface. A thematic raster, however, represents a category or quantity of a phenomenon such as elevation, pollution, population, rainfall, or noise. Since readings cannot be taken at every location, samples are taken instead, and a surface model is made. The model approximates the surface by interpolating the values between the sample points.

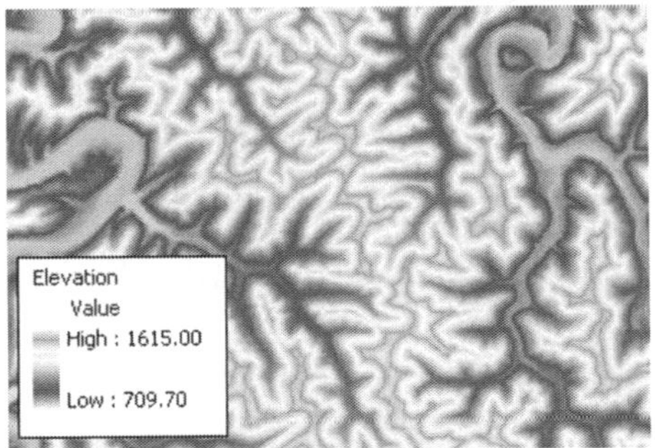

A thematic raster of elevation values. A few of the cells represent samples actually taken, but most of the values have been interpolated

3D Analyst uses the z value stored in each cell to display the raster in 3D. Elevation values are commonly shown, but any numeric cell value can be illustrated in three dimensions. Even though images and many thematic rasters don't contain elevation values, you can still display them in 3D by draping them over a 3D surface model with the same geographic extent.

TINs

A Triangulated Irregular Network (TIN) represents a surface as a set of irregularly located points, joined by lines to form a network of contiguous, non-overlapping triangles that vary in size and proportion. Each triangle node stores an x, y, and z value.

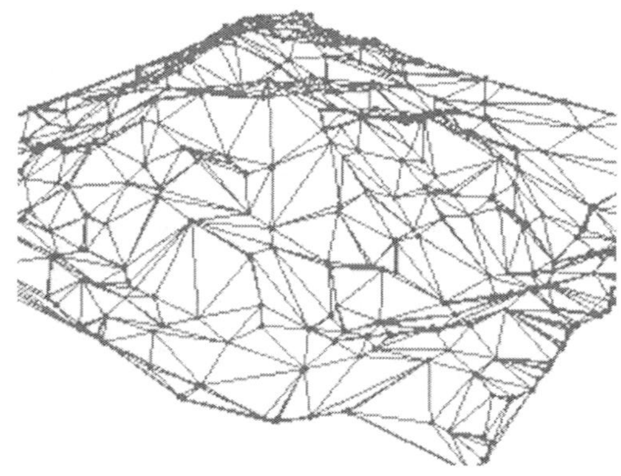

The structure of a TIN. Top: only the TIN edges and nodes are shown. Bottom: the TIN's triangles (faces) are colored to represent elevation

Like rasters, the values in a TIN are interpolated from sample points. The sample points form the triangle nodes, and the interpolation (or triangulation, as it's generally called) consists of connecting the nodes by lines. Once the TIN is built, the elevation of any location on a TIN surface can be estimated using the x, y, and z values of the bounding triangle's vertices. The slope and aspect for each triangle face is also calculated.

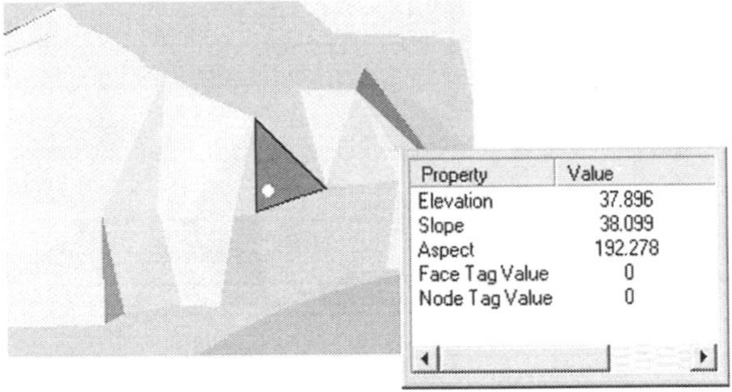

When you identify any point on the face of a TIN, the node x, y, z values are used to interpolate the elevation at that point. The node values are also used to calculate the slope and aspect of each triangle face

Because the nodes can be placed irregularly over the surface, TINs can show greater detail where a surface is highly varied or where you want more accuracy. A TIN is only as good as the initial sample points taken; mountainous areas need many more samples per square unit than flat areas do in order to create an accurate terrain model.

TIN models are less widely available than raster surface models; they take longer to build and require much more disk space. They are typically used for precise modeling of small areas.

Terrain Datasets

A terrain dataset represents a surface in the same way that a TIN does, but uses a different storage system. Elevation measurements collected by LiDAR or SONAR are typically used to create a terrain because they can result in millions of mass points. This enables the creation of a very accurate surface, but also makes for a large file size. Terrain datasets get around this by separating the TIN representations into multiple levels of resolution, called pyramids, for faster drawing of large datasets. At smaller scales, a coarser TIN representation is drawn; at larger scales, more detailed features are shown.

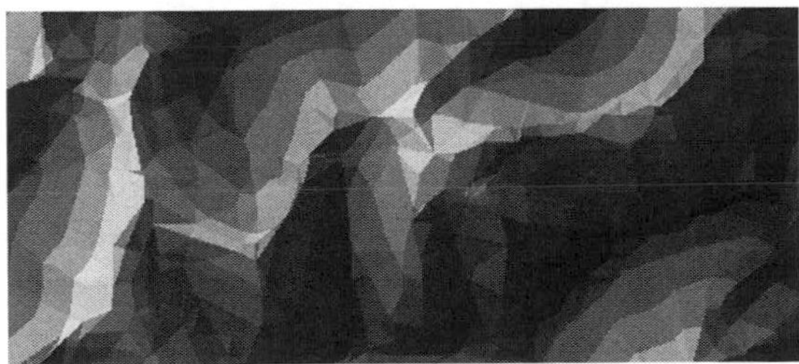

In a lower-resolution pyramid of a terrain dataset, fewer triangles are calculated during display. This pyramid level would normally be drawn at a smaller scale in ArcMap

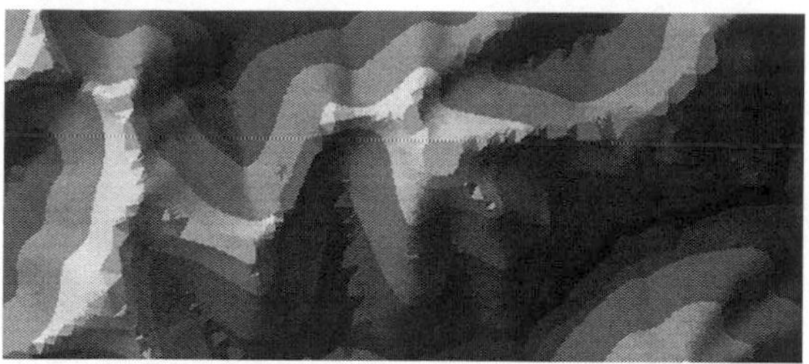

In the finest resolution pyramid, all triangulations are represented in the terrain dataset. This pyramid level would normally be drawn at a large scale in ArcMap

3D Features

3D vector features, like their 2D counterparts, represent objects or clearly bordered areas such as buildings, land parcels, roads, power poles, and wells. Often, the z values in 3D features are used to represent an attribute other than height. For example, you might create a scene that shows city points extruded into 3D columns based on their population.

Like TINs, 3D features store z values along with x, y coordinates as part of their geometry. A point has one z value; lines and polygons have one z value for each vertex in the shape. You can identify 3D feature classes by looking at the Shape field in their attribute tables.

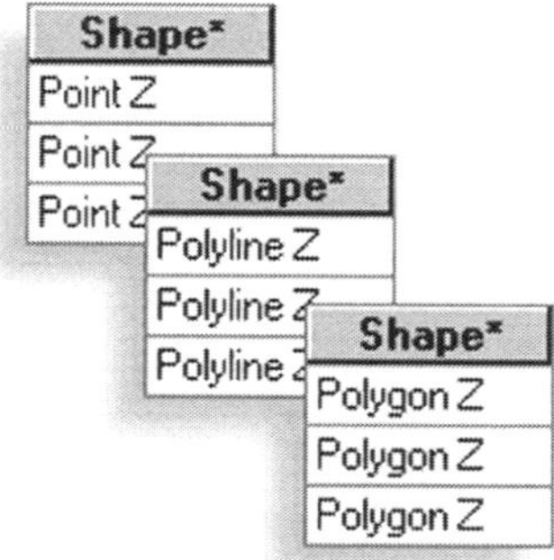

A 3D feature class shows a Z value in its Shape field

2.5-Dimensional and 3-Dimensional Data

Generally, when we talk about 3D data, we mean 2.5 D data. Rasters, TINs, and terrains are surfaces that store exactly one z value for each x, y value pair. Also, when we turn points, lines, or polygons into solids, we're really just extruding 2D coordinates by a set of specified values.

True 3D geometric structures are represented in 3D Analyst by multipatch features. Multipatches are made of planar 3D rings and triangles stitched together to model objects like spheres, trees, rooftops, or buildings with overhanging features.

Software programs like Sketchup (Google's free 3D modeling software), 3ds Max, OpenFlight, and VRML 2.0 can create models that 3D Analyst can import into a geodatabase and use as symbols. They can also be used as graphics in ArcScene or ArcGlobe, without committing them into a geodatabase.

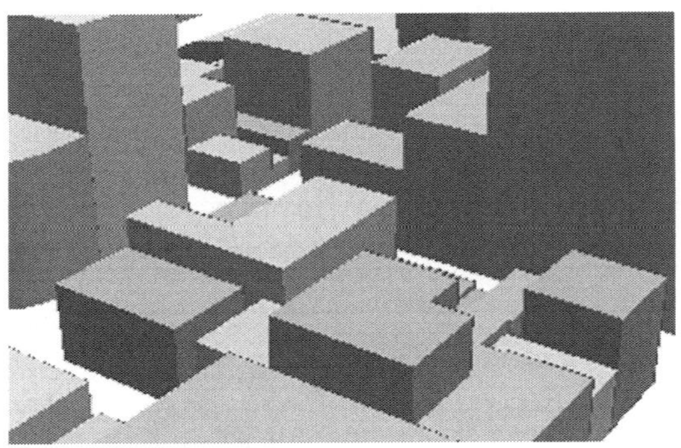

3D features extruded from 2D polygons. Each x,y vertex has one companion z value

Multipatch (true 3D) features can handle multiple z values per x,y vertex, permitting realistic representations of features that include overhangs or textures

KML

KML is Google Earth's file format for displaying 3D data. You may be used to seeing the most common form of KML document in Google Earth and Google Maps: the Placemark. This is similar to a point feature class in ArcGIS in that it represents a geographically referenced point on the earth. It differs completely, of course, in the file structure and instructions used for rendering. KML is a tag-based structure similar to HTML and XML that can be created or altered in any text editor. Generally, though, you will create 3D features directly in Google Earth, or import files created in Sketchup into Google Earth.

Besides Placemarks, KML allows you to create or place polygons, lines, and raster image overlays in Google Earth.

A Placemark and a Polygon created in Google Earth. The KML code for the Placemark object is shown below:

```
<name>KmlFile</name>
        <Style id="sn_ylw-pushpin_copy0">
                <IconStyle>
                        <color>ff0055ff</color>
                        <scale>1.7</scale>
                        <Icon>
                                <href>http://maps.google.com/mapfiles/kml/pushpin/ylw-pushpin.png</href>
                        </Icon>
                        <hotSpot x="20" y="2" xunits="pixels" yunits="pixels"/>
                </IconStyle>
                <LabelStyle>
                        <color>ff000000</color>
                </LabelStyle>
        </Style>
        <StyleMap id="msn_ylw-pushpin_copy0">
                <Pair>
                        <key>normal</key>
                        <styleUrl>#sn_ylw-pushpin_copy0</styleUrl>
                </Pair>
                <Pair>
                        <key>highlight</key>
                        <styleUrl>#sh_ylw-pushpin_copy0</styleUrl>
                </Pair>
        </StyleMap>
        <Style id="sh_ylw-pushpin_copy0">
                <IconStyle>
                        <color>ff0055ff</color>
                        <scale>2.00909</scale>
                        <Icon>
                                <href>http://maps.google.com/mapfiles/kml/pushpin/ylw-pushpin.png</href>
                        </Icon>
                        <hotSpot x="20" y="2" xunits="pixels" yunits="pixels"/>
                </IconStyle>
                <LabelStyle>
                        <color>ff000000</color>
                </LabelStyle>
        </Style>
        <Placemark>
                <name>Untitled Placemark</name>
                <LookAt>
                        <longitude>-95.26548319412244</longitude>
                        <latitude>38.95938957105109</latitude>
                        <altitude>0</altitude>
                        <range>11001000</range>
                        <tilt>0</tilt>
                        <heading>7.256941233334342e-015</heading>
                        <altitudeMode>relativeToGround</altitudeMode>
                </LookAt>
                <styleUrl>#msn_ylw-pushpin_copy0</styleUrl>
                <Point>
                        <coordinates>-95.26548319412244,38.9593895710511,0</coordinates>
                </Point>
        </Placemark>
</Document>
</kml>
```

Now that you've had an introduction to 3D data structures, you're ready to learn how to use 3D Analyst.

Load the Tutorial Data

The next two exercises will teach you how to work with 3D data in ArcCatalog. Before you can do any exercises, however, you need to load the 3D Analyst tutorial data and add the ArcScene, ArcGlobe, ArcCatalog, and ArcMap program icons to your desktop.

Visit the support website for the book to download the tutorial data at www.wiley.com/college/kennedy. Click on the cover for *Introduction to 3D Data: Modeling with ArcGIS 3D Analyst and Google Earth*. On the next page click the link for the Student Companion Site on the right side of the page and follow the links for the 3DDATA.zip archive. Download the zip file to your desktop and extract the archive to the drive of your choice using a program such as WinZip or 7-Zip (freely available at www.7-zip.org). If you have the disk space, I recommend that you copy it directly under your C: drive, so that the full pathname reads "C:\3DDATA."

Once the contents of the 3DDATA.zip archive are copied to your hard drive, you can delete the zip file as you won't need it again.

Add the Program Icons to Your Desktop

Note: ArcGIS 3D Analyst must be installed before you can create shortcuts to it on your desktop. If you have not installed the software, please see the ArcGIS 3D Analyst installation guide.

1. On the taskbar of your desktop, click the Start menu. Move your cursor to Programs, then ArcGIS, and right-click on ArcScene.
2. Choose Send to . . . , and then click Desktop (Create Shortcut). A shortcut to ArcScene is added to your desktop. (This procedure is for Windows XP; if you're running Windows 2000, NT, or Vista, it may be a little different.)
3. Use the same procedure to add the ArcGlobe icon to your desktop. If you don't already have desktop icons for ArcCatalog and ArcMap, you should add them as well.

Exercise 1-1

Preview Data in ArcCatalog

With 3D Analyst loaded in ArcCatalog, you can preview both 2D and 3D data in three dimensions. In this exercise, you'll examine a TIN of Coletown, KY, and a 3D shapefile of contour lines.

Step 1. Start ArcCatalog

Double-click the ArcCatalog icon on your desktop. If you didn't make an ArcCatalog desktop icon, either see the instructions immediately above, or click the Start menu, point to Programs, point to ArcGIS, and click ArcCatalog.

From the Tools menu, choose Options.

Click the General tab. At the bottom of the dialog, uncheck the box next to "Hide file extensions."

Click OK.

Step 2. Load the 3D Analyst Extension

From the Tools menu, choose Extensions. In the dialog, check the box next to 3D Analyst. Click Close.

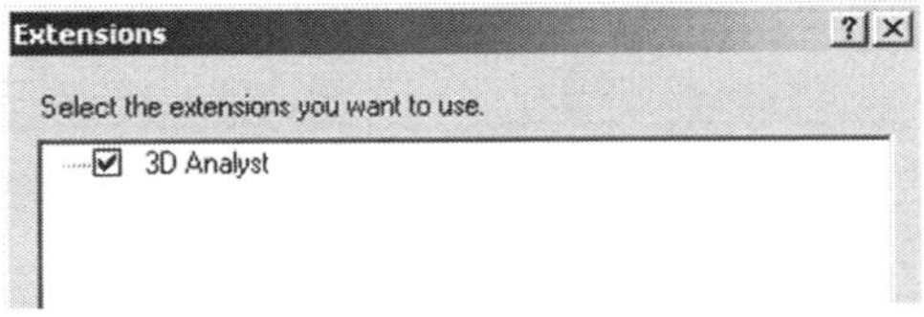

Step 3. Load the 3D View Toolbar in ArcCatalog

From the View menu, click Toolbars and check the box next to 3D View Tools and Globe View Tools.

These tools let you view and navigate your data in 3D and in Globe view, query 3D features, and create perspective-view thumbnails.

Other than the Launch ArcScene button, 

the Launch ArcGlobe button, 

and the Create Thumbnail buttons, 

these tools are also found on the Tools toolbars in ArcScene and ArcGlobe, respectively.

Step 4. Look at ArcCatalog's Contents

In the Catalog tree on the left, navigate to the 3DDATA\Chapter01\Data folder. Click the plus sign next to the Data folder to open it.

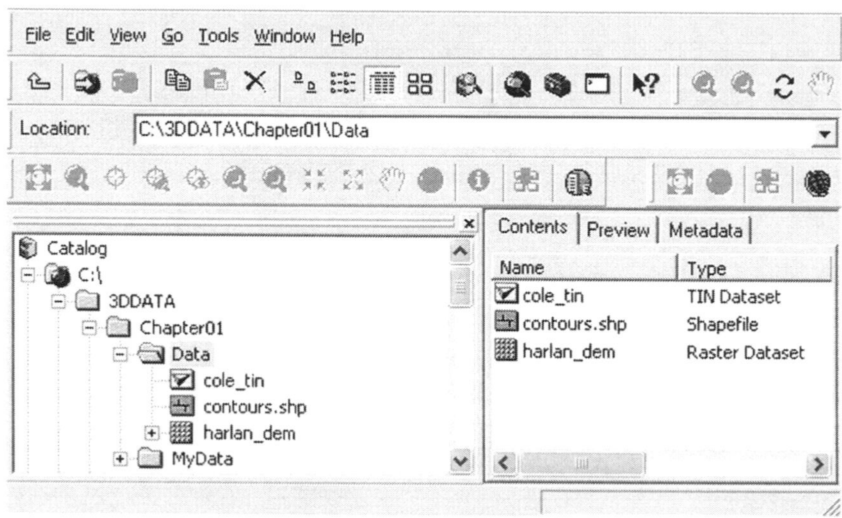

The 3DDATA\Chapter01\Data folder contains one TIN, one raster, and one shapefile feature class. Notice the icons that ArcCatalog uses to represent TIN and raster data types.

Step 5. Preview the TIN Dataset

Click cole_tin in the Catalog tree, then click the Preview tab above the display. The TIN is displayed in two dimensions (also called the orthographic or planimetric view). The faces of the TIN are colored by elevation value.

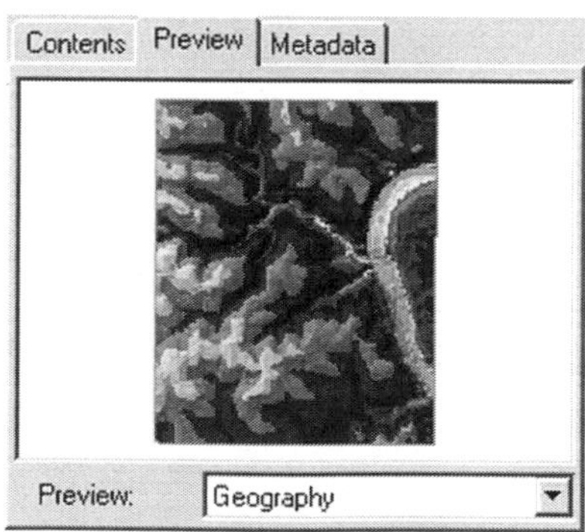

Click the arrow next to the Preview menu below the display, and select 3D View.

The TIN is displayed in three dimensions (perspective view). By default, the angle is from the southwest, and the TIN is no longer symbolized by elevation values but by a single color for all faces. This was introduced in version 9.0 as an attempt to speed up drawing time.

Step 6. Use the Navigate, Zoom, and Pan Tools

The three tools that you will use most to examine your 3D data are Navigate, 

Zoom In/Out, 

and Pan. 

The Navigate tool lets you rotate your data in any direction around the center of the display. Click the Navigate tool.

Place your cursor over the center of the TIN in the display, hold down the left mouse button, and move the mouse in any direction. You can inspect the TIN from any angle (except from underneath; for that you need to be in ArcScene).

To reset the view, click the Full Extent button. 

Click the Zoom In/Out tool. Place your cursor at the top center of the TIN in the display. As in ArcMap, holding down the left mouse button while dragging the mouse up and down zooms the display in and out.

Click Full Extent again.

Not surprisingly, the Pan tool moves your data horizontally, vertically, or diagonally across the display. Click the Pan tool. Place your cursor over the center of the display, hold down the left mouse button and drag the cursor in any direction. When you're finished, click Full Extent.

Step 7. Pan and Zoom with the Navigate Tool

In addition to rotating your data in the display, the Navigate tool also lets you pan and zoom without changing tools.

Click the Navigate tool again. 

Place the cursor over the center of the display. Hold down the center mouse wheel and drag the cursor in any direction. When you're comfortable panning with the Navigate tool, click Full Extent.

With the Navigate tool still selected, hold down the right mouse button. Drag it down in the display to zoom in, up to zoom out. You can also scroll with the mouse wheel to achieve the same effect. When you're comfortable zooming with the Navigate tool, click Full Extent.

Step 8. Experiment with the Other Navigation Tools

The Narrow and Expand Field of View buttons and the Zoom In and Zoom Out buttons work the same way in ArcScene as they do in ArcMap.

Feel free to experiment with them. When you're finished, set the display back to Full Extent.

Step 9. Preview 3D Vector Features

In the 3DDATA\Chapter01\Data folder in the Catalog tree, click contours.shp. Make sure the Preview tab above the display is selected. Click the Navigate tool, and set the Preview menu below the display to 3D View.

(If you only see "contours," not "contours.shp" in the Data folder, click the Tools menu in ArcCatalog and choose Options. On the General tab, uncheck the box at the bottom that says "Hide File Extensions.")

Put your cursor over the contour lines in the Preview display, hold down the left mouse button and rotate the contours in different directions. Hold down the right mouse button to zoom in, then navigate some more.

Notice that the contour data rotates more quickly and smoothly than the TIN data. That's because the contour shapefile takes up much less disk space.

Step 10. Compare the File Sizes of the Datasets

Click on the Metadata tab above the display. Enlarge the window so you can see all three tabs: Description, Spatial, and Attributes.

Click on Description. Scroll down to "Data storage and access information," and click on it. Scroll down some more, and click on "Accessing the data." Note that the file size of contours.shp is 0.070 MB.

In the Catalog tree to the left of the Preview display, click on cole_tin. Make sure that the Metadata tab and the Description heading are selected. Scroll down again to "Data storage and access information," click on it, and scroll further to "Accessing the data."

The file size of cole_tin is 0.429 MB, six times larger than contours.shp. (Depending on a few factors having to do with metadata and thumbnail graphics, your file sizes may vary somewhat.)

Step 11. Preview the 3D Shapefile Attribute Table

As you read earlier, one way to tell if a feature class contains 3D features is to look at its attribute table. You can do this in ArcScene, ArcGlobe, ArcMap, or ArcCatalog. In the Catalog tree, click on contours.shp again. Click on the Preview tab above the display, and set the Preview menu below the display to Table.

Contents | Preview | Metadata

FID	Shape*	CONTOUR
0	Polyline ZM	146
1	Polyline ZM	144
2	Polyline ZM	142
3	Polyline ZM	140
4	Polyline ZM	138
5	Polyline ZM	136
6	Polyline ZM	134
7	Polyline ZM	132
8	Polyline ZM	130
9	Polyline ZM	128

Record: 1 Show: All Selected

Preview: Table

Look at the Shape field in the table. The xpression *Polyline ZM* indicates that this is a 3D polyline feature. A 2D polyline feature would have a Shape field value of *Polyline.*

Step 12. Query 3D Shapefile Attributes

You can also select individual 3D features and look at the attributes in ArcCatalog. In the Preview menu below the display, select 3D View. Notice that all of the 3D tools become active again.

Click the Navigate tool. Put your cursor over the display, hold down the right mouse button, and drag it downward to zoom in on the contour lines.

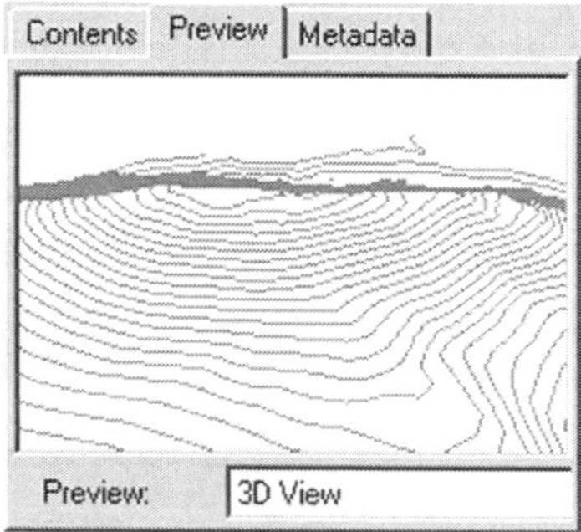

Click the Identify tool.

Click on one of the contour lines. ArcCatalog highlights it and the Identify Results box appears, listing the contour line's ID number, its z-value (height), and the x,y location at the point clicked.

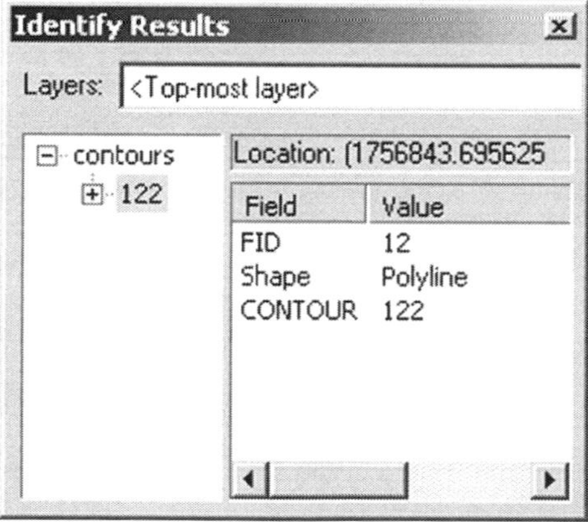

Click on a few more contour lines, and note their elevations. When you're finished, close ArcCatalog.

In this exercise, you loaded the 3D Analyst extension in ArcCatalog, previewed a TIN and a 3D shapefile, learned to use the 3D navigation tools, looked at 3D feature class attributes, and examined metadata. In the next exercise, you'll preview a raster dataset in ArcCatalog and create a layer from it.

Exercise 1-2

Create a Layer File in ArcCatalog

In this exercise you'll preview a Digital Elevation Model of Harlan, KY, create a layer file, symbolize it, and make a 3D thumbnail.

Step 1. Start ArcCatalog

Double-click the ArcCatalog icon on your desktop. If you didn't make an ArcCatalog icon, click the Start menu, choose Programs, then ArcGIS, then ArcCatalog.

Step 2. Preview the Raster Dataset

Navigate to the contents of your 3DDATA\Chapter01\Data folder, and click on harlan_dem.

Click the Preview tab above the display.

Change the Preview menu from Geography to Globe View, and then to 3D View. Notice that even though the DEM is shown in 3D perspective, it still looks flat.

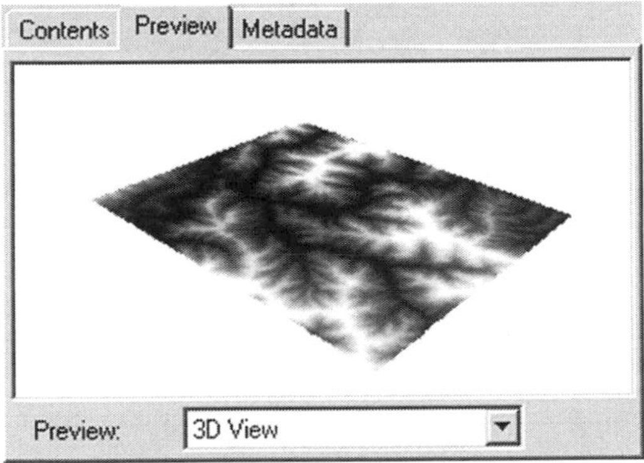

As you saw in Exercise 1, TINs and 3D vector features display their height values in ArcCatalog when you select 3D View. Rasters, however, are drawn as though they lie on a flat surface. In order to see the heights of an elevation raster in ArcCatalog, you have to create a layer file (.lyr) from the raster and specify its 3D drawing properties.

Step 3. Create a Layer File from the Raster

In the Catalog tree, right-click harlan_dem and choose Create Layer. Name it harlan_layer, and save it in your 3DDATA\Chapter01\MyData folder (not the Data folder).

Again in the Catalog tree, open the MyData folder and click once on harlan_layer.lyr to highlight it.

Select the Preview tab above the display, and choose 3D View from the Preview menu.

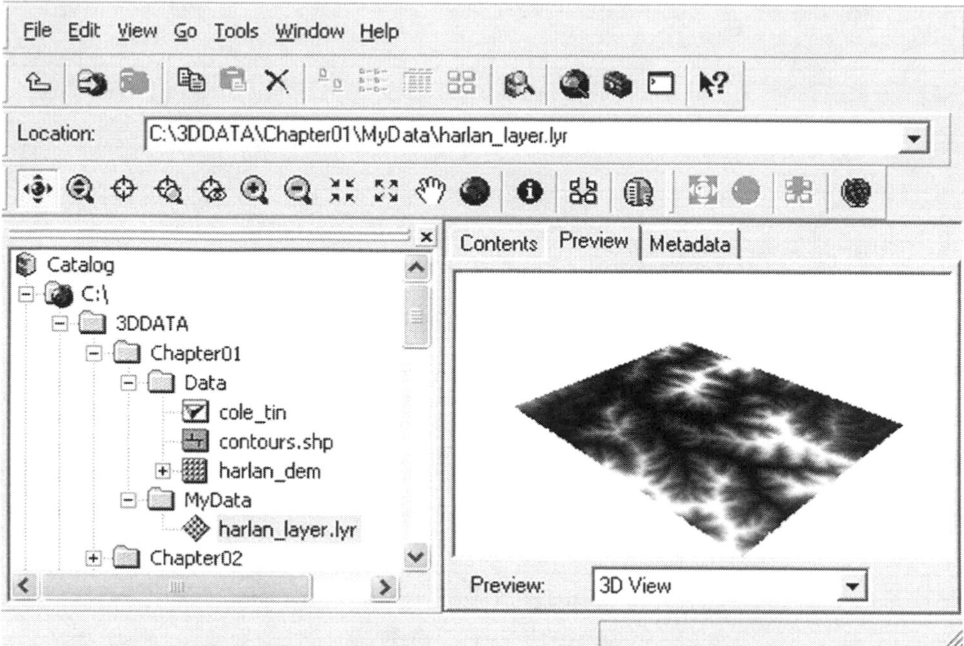

Harlan_layer .lyr is not a copy of harlan_dem; in fact, it's not a raster dataset at all. As with the .lyr files for other data types that you may have used previously with ArcCatalog, it's a much smaller file that contains a copy of the display instructions for harlan_dem. You can't change the 3D viewing properties of the original harlan_dem in ArcCatalog, but you can change the 3D viewing properties of harlan_layer.

Step 4. Set Base Heights for Harlan_layer

In the Catalog tree, right-click harlan_layer and click Properties.

In Layer Properties, click the Base Heights tab. Choose "Obtain heights for layer from surface," and make sure that the surface used is harlan_dem from the Chapter 1 Data folder.

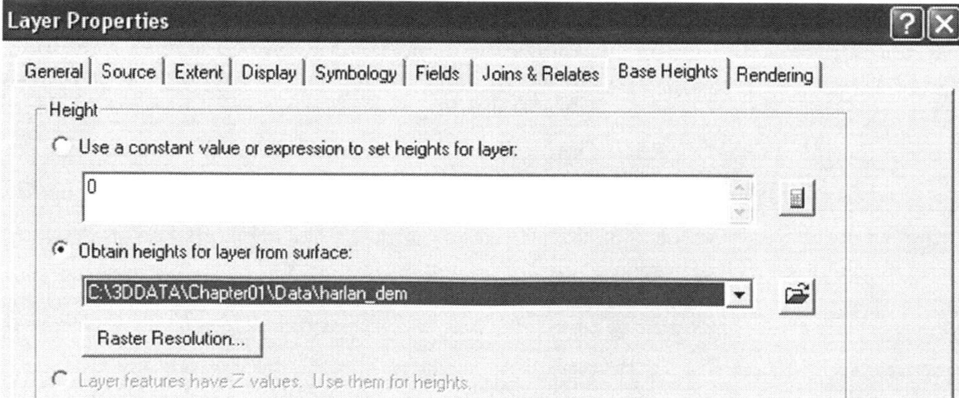

This setting uses the elevation values stored in harlan_dem to define the base heights of harlan_layer. You'll learn more about base heights in the Chapter 2.

Click OK to close the Layer Properties dialog, and look at harlan_layer in 3D View.

Harlan_layer draws in 3D, but you'll change the color scheme to better reveal its elevation levels.

Step 5. Change the Layer's Color Scheme

Right-click harlan_layer in the Catalog tree again, and click Properties.

In Layer Properties, click the Symbology tab. In the Color Ramp dropdown list, right-click on the color ramp itself (not on the dropdown arrow). Click Graphic View to uncheck it. This replaces the color ramp with its name.

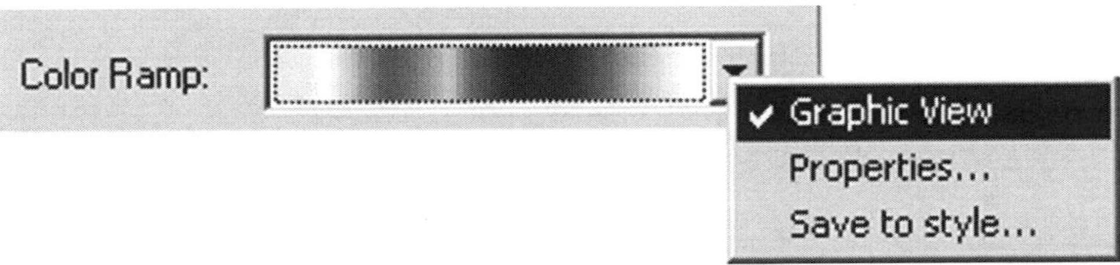

Click the Color Ramp dropdown arrow and scroll down until you see Elevation #1. Click to select it.

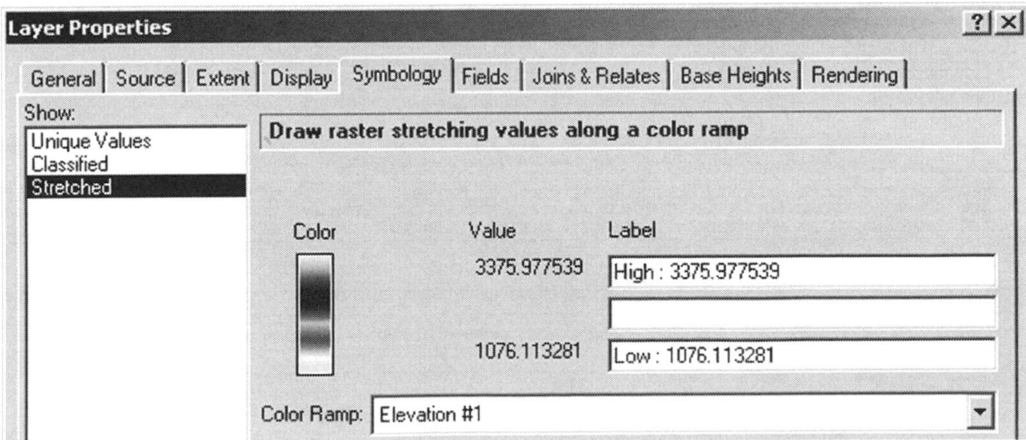

Click OK to close the Layer Properties dialog.

Preview harlan_layer again in 3D.

The elevation levels of this piece Harlan County are much more apparent now, but you can improve the perspective further by adding shading to the surface.

Step 6. Add Shading to the Layer

Open Layer Properties again for harlan_layer.

Click the Rendering tab. In the Effects frame, check "Shade areal features relative to the scene's light position."

Click OK.

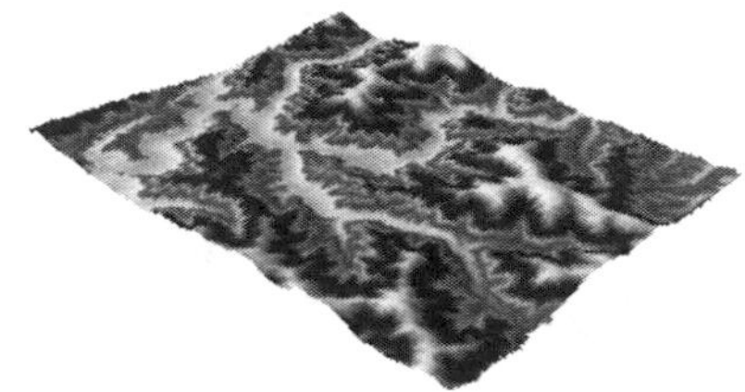

Because you illuminated the surface from a realistic sun angle, the elevation of the landscape stands out in greater relief. You'll learn more about illumination in Chapter 2.

Step 7. Create a 3D Thumbnail

In the Catalog tree, click once on harlan_layer.lyr, and select the Contents tab above the display.

ArcCatalog shows the name, file type, and file size of harlan_layer. You'll create a 3D thumbnail to go along with the information.

Click on the Preview tab above the display, and select 3D View from the Preview menu. The 3D tools are activated.

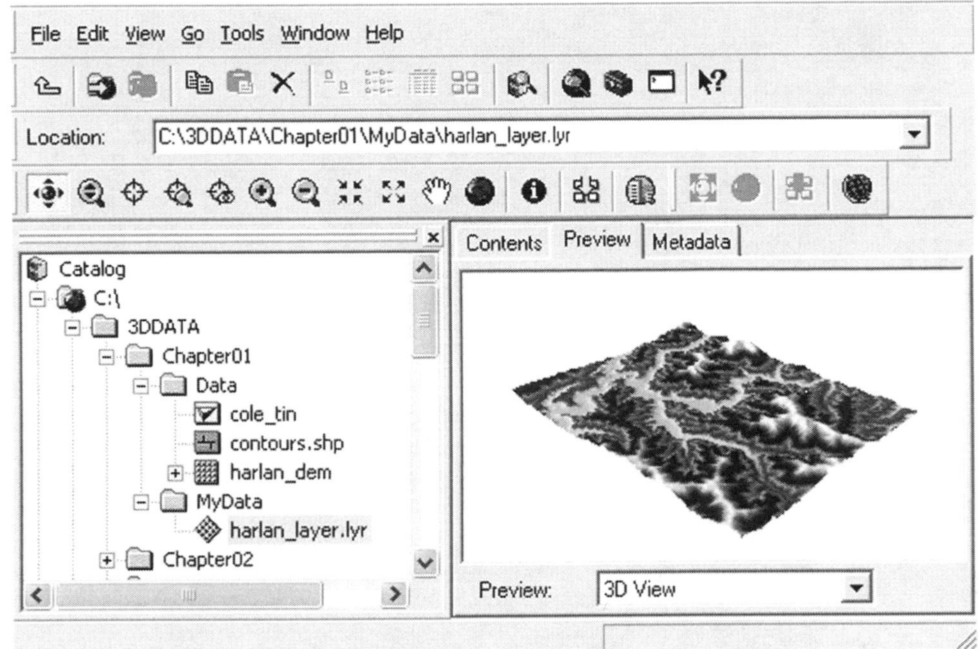

Click the Create Thumbnail button on the 3D toolbar.

Click the Contents tab again. The new harlan_layer 3D thumbnail is added to the file description.

Close ArcCatalog.

In this chapter, you learned how to use the 3D Analyst navigation tools and how to preview, create, and symbolize 3D data in ArcCatalog. Chapter 2 will introduce you to the ArcScene interface.

CHAPTER **2**

3D Display in ArcScene

Now that you know a bit about the types of data you can use in 3D Analyst, let's take a look at how ArcScene draws the data in 3D. ArcScene operates very much like ArcMap; you can add multiple data layers to a scene such as rasters, TINs, 2D, or 3D feature classes. ArcScene, however, lets you control how each layer is symbolized and positioned in 3D space. You can also change general settings that apply to the entire scene, such as sun illumination, background color, and vertical exaggeration.

Scene Properties: Background Color, Illumination, Coordinate System, and Vertical Exaggeration

Background Color

By default, the background color in ArcScene is white. You can change it to whatever color you like, such as blue or black to simulate earth sky or pinkish-orange to suggest the dust on Mars. You can choose colors from the Scene Properties palette or mix your own. You can also set a default background color for all new scenes.

Illumination

Every 3D scene has a light source that determines which parts of the surface are illuminated and which are in shadow. The position of the light source is controlled by its altitude and azimuth settings.

The altitude is the angle, measured in degrees from 0 to 90, between the light source and the horizon. An altitude of 0 degrees puts the light source level with the horizon; an altitude of 90 degrees puts it directly overhead. ArcScene places the default altitude at 30 degrees.

The azimuth is the compass direction from which the sun or other light source shines on the scene. It's measured clockwise in degrees from 0 (due north) to 360 (also due north). The default azimuth setting for the light source is 315 degrees, placing it in the northwest.

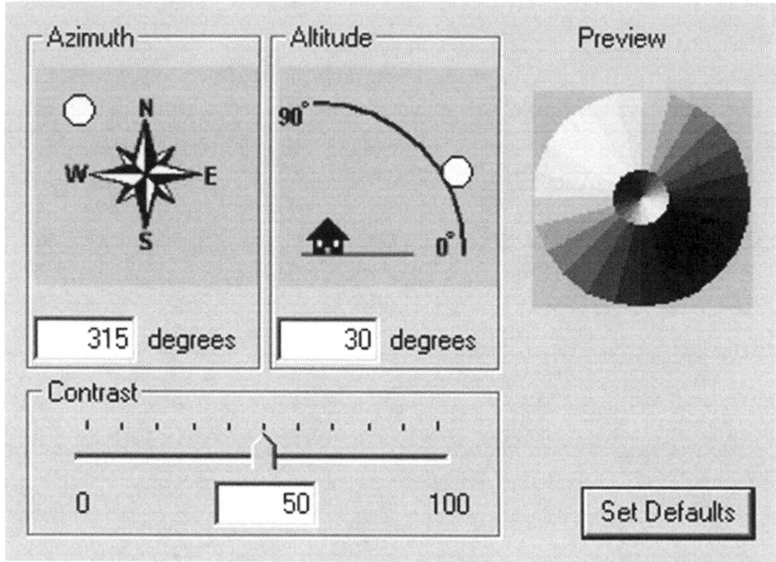

The default light source settings in ArcScene

You can change the azimuth and altitude of the light source and alter the level of contrast. The illumination will affect all the surfaces and vector features in a scene.

Coordinate System

When you open an empty scene, it has no coordinate system and adopts the coordinate system of the first data layer you add to it. This coordinate system overrides those of any subsequent layers added to the scene. If you add more layers that are in other systems, they are temporarily altered to match the one in the scene. This is called on-the-fly projection, and puts layers in different coordinate systems into spatial alignment.

You can select a different coordinate system for the scene at any time, and all layers in the scene will be temporarily reprojected. This will not alter the coordinate systems of the source datasets.

Like ArcMap, ArcScene expects coordinate system information to be stored with each data source. If it isn't, you'll need to know what system the dataset uses so that you can give ArcScene the information. For a layer in a geodatabase, this information is part of the layer's metadata. For coverages, shapefiles, TINs, and rasters, it's stored in a separate file named after the dataset, but with a .prj file extension (for example, parcels.prj). If the .prj file hasn't been provided with the dataset but you have the coordinate system information, you can create one with ArcCatalog.

If no coordinate information is associated with a dataset, ArcScene will test the coordinate values to see if they fall within the range of −180 to 180 degrees for x-values and −90 to 90 degrees for y-values. If they do, ArcScene assumes that these are geographic coordinates of latitude and longitude. If the values aren't in this range, ArcScene considers them to be planar x,y coordinates.

Vertical Exaggeration

You can exaggerate the vertical appearance of surfaces by multiplying the z units in a scene by a number. A vertical exaggeration of 2 multiplies all z values by 2, an exaggeration of 0.4 multiplies all z values by 0.4, and so on. It's often used to emphasize slight changes in elevation on a surface that looks flat because of its large extent. Conversely, a fractional vertical exaggeration can smooth choppy surfaces.

A TIN with no vertical exaggeration

The same TIN with a vertical exaggeration of 3 applied to the scene

Vertical exaggeration is also used to bring z values into proportion with x,y values when their units are different. For example, the x,y values of a raster may be in Universal Transverse Mercator (UTM) meters but the elevation measurements may have been taken in feet.

Like a scene's illumination and coordinate system, vertical exaggeration affects all the layers in a scene. It is just a visual effect, and doesn't influence measurement or analysis.

3D Layer Display Properties: Base Heights, Extrusion, and Shading

In addition to the global settings that affect all layers in a scene, each layer has its own set of 3D drawing properties.

Base Heights

The base height of a layer is the elevation at which it is drawn in a scene. As you saw when you loaded 3D Analyst in ArcCatalog, TINs and 3D features are automatically drawn in 3D, but rasters and 2D features look flat. The same holds for ArcScene. ArcScene gets the base heights for TINs and 3D features from the z values in their nodes and vertices, respectively. In order to draw rasters in 3D, however, you have to tell ArcScene where to get the z values to use for base heights. ArcScene doesn't automatically use the values in the cells of a raster for base heights because quite often rasters model something other than elevation, for example in an aerial photo or a grid of pollution dispersion. (Non-elevation rasters and 2D vector features are instead "draped" over another 3D surface, such as an elevation raster or a TIN, whose z values are used for base heights.)

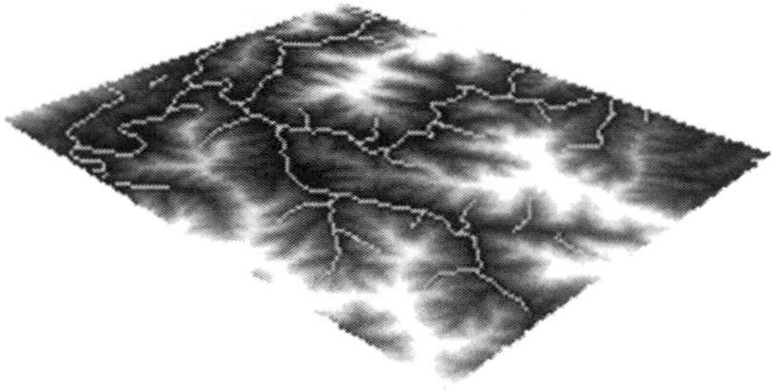

An elevation raster and a 2D feature layer of rivers. When they're first added to ArcScene their base heights are set to 0, so they lie flat

The elevation raster's base heights have been set to its cell values, so ArcScene displays it in 3D. Base heights have not yet been set for the features, so they lie underneath the raster

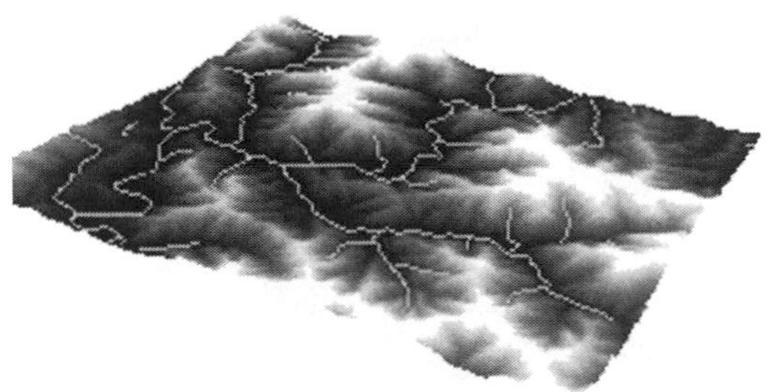

The base heights for the 2D river features have been set to the base heights of the raster. Now the rivers drape over the surface

Extrusion

Extrusion extends 2D vector features vertically from their base height to a point that you specify. You can extrude points into vertical lines, lines into walls, and polygons into solids.

2D polygons become blocks when extruded into three dimensions

Shading

You can also control whether or not individual layers are shaded by the illumination settings for the scene. Shading gives layers a more realistic appearance. While the Illumination setting is naturally the same for all layers in a scene (the light source comes from one altitude and one angle) whether a layer "participates" in the shading is up to you.

An unshaded elevation raster

The same raster, shaded according to the sun's altitude and azimuth settings in the scene

In this chapter you'll work with ArcScene. You'll learn to set background color, illumination, vertical exaggeration, and coordinate system; you'll set base heights for an elevation raster, an aerial photo, and 2D vector features; and you'll extrude 3D polygon features into solid shapes.

Exercise 2-1

Set Background Color and Illumination in ArcScene

When you open a new document in ArcScene, 3D Analyst places the sun in the northwest at 30 degrees above the horizon, so that north and west-facing surface slopes are lit and south and east-facing slopes are in shadow. In this exercise, you'll learn how to change these settings.

Step 1. Start ArcScene

Double-click the ArcScene icon on your desktop. If you didn't make an ArcScene icon in Chapter 1, click the Start menu, point to Programs, point to ArcGIS, and click ArcScene.

Step 2. Load the 3D Analyst Toolbar

In ArcScene, the 3D Analyst extension should be loaded by default. Check this by clicking on the Tools menu, then Extensions. The box next to 3D Analyst should be checked.

From the View menu, choose Toolbars, then check the box next to 3D Analyst. This loads the 3D Analyst menu and toolbar.

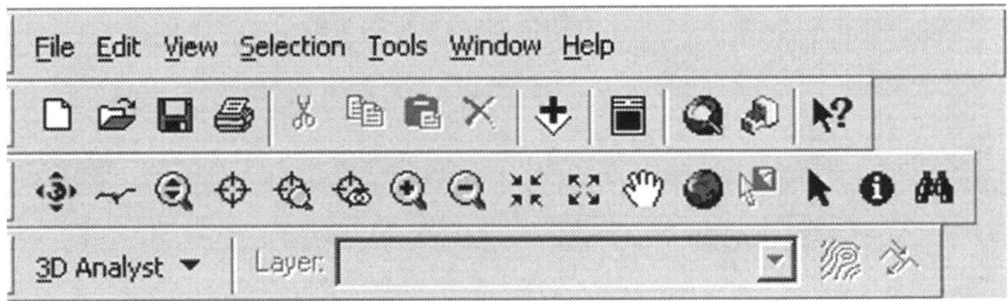

Step 3. Add Hill_tin.lyr to the Scene

Click the Add Data button in ArcScene. 

Navigate to your 3DDATA\Chapter02\Data folder. Double-click hill_tin.lyr to add it to the scene.

Hill_tin.lyr is a layer file that uses the TIN surface hill_tin for its source. Hill_tin is also in your 3DDATA\Chapter02\Data folder. (Note: Hill_tin.lyr was made for ArcScene version 9.3. If you are running version 9.1 or 9.2, add hill_tin_v91.lyr to ArcScene instead. It also uses hill_tin as its data source.)

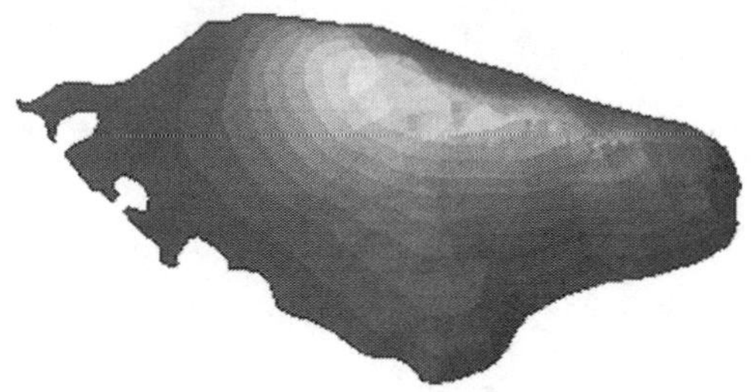

Click the Navigate tool, 

and hold down the left mouse button while moving the cursor around in the scene. Hold down the right mouse button, and move the cursor downward in the scene to zoom in to the hill.

Step 4. Set a Background Color for the Scene

In the Table of Contents, right-click Scene Layers and choose Scene Properties. Click the General tab.

Click the Background Color dropdown arrow, and choose Sodalite Blue.

Click OK.

The scene's background color changes to a light blue.

Step 5. Add a Navigation Marker

Click the Full Extent button, then click the Add Data button again.

Open your 3DDATA\Chapter02\Data folder. Double-click south_end.lyr to add it to the scene. (Again, if you are running version 9.1 or 9.2, add south_end_v91.lyr to the scene instead.)

The yellow marker sits on the southern slope of the hill to help you keep your bearings as you navigate. Notice that your default vantage point at full extent is from the southwest.

Step 6. Change the Level of Contrast in the Scene

Click the Navigate tool again.

Hold down the left mouse button in the scene, and move the surface from right to left until you are looking at the hill from the south instead of the southwest. Hold down the right mouse button to zoom in, and hold down both mouse buttons together to pan. The scene should look something like this:

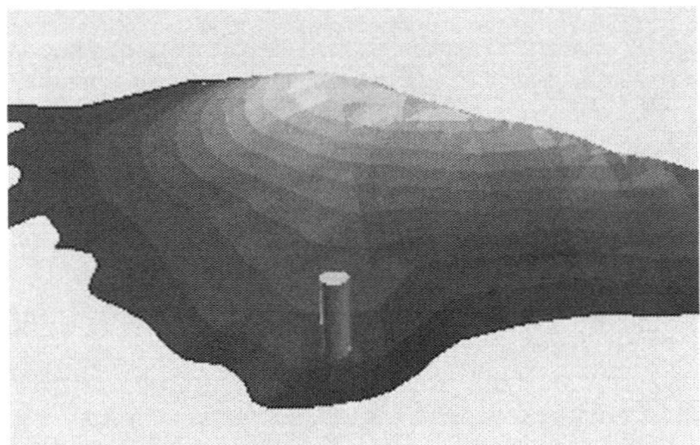

Move the ArcScene application window to one side of your computer screen.

In the Table of Contents, right-click Scene Layers and choose Scene Properties. (You can also double-click the words "Scene Layers" to bring up the Scene Properties.) Move the Scene Properties dialog so that you can see the dialog box and the scene.

Click the Illumination tab.

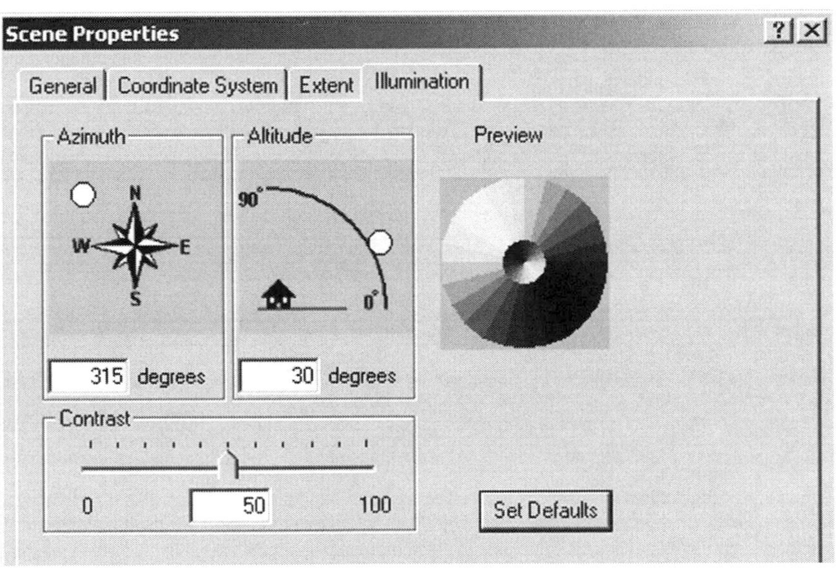

Drag the Contrast slider bar back and forth between 0 and 100. The slider bar controls the amount of shading applied to the surface. As the contrast is increased, the shadows deepen, emphasizing variation on the surface.

The default Contrast setting is 50. In the Contrast box, highlight the value and type 75. Leave the Scene Properties dialog open.

Step 7. Change the Sun's Altitude and Azimuth

You can change the sun's position by moving the sun icon with your cursor or by typing in values.

In the Azimuth frame, drag the sun icon from the northwest to the west, then south, then east. Alternating between lighting the northern and southern sides of the hill will give you a good idea of how azimuth is modeled in ArcScene.

Put the sun in the northwest in the Azimuth frame.

In the Altitude frame, drag the sun icon up to 90 degrees and then down to zero. The higher the sun, the more directly each face of the TIN's surface is lit. At 90 degrees (midday) most of the detail of the hill is obliterated. At 0 degrees (sunset) it is almost completely in shadow.

Step 8. Experiment with Illumination Settings

Click Restore Defaults to restore the original settings. Feel free to play with the Azimuth, Altitude, and Contrast settings.

If you want to navigate around the scene, you have to click OK to close the Scene Properties dialog first. To reopen the dialog, right-click Scene Layers in the Table of Contents.

Step 9. Close ArcScene

Close the Scene Properties dialog. If you'd like to save the ArcScene document, go to the ArcScene File menu and click Save. Give it a name of your choice, and save it in your 3DDATA\Chapter02\MyData folder (not the Data folder). The .sxd extension indicates that the file is an ArcScene document, just as an .mxd extension indicates an ArcMap document.

Close ArcScene.

Exercise 2-2

Set Vertical Exaggeration in ArcScene

Vertical exaggeration is used to emphasize small changes in elevation, or to bring z values into proportion with x,y values when they're in different units. In this exercise, you will apply vertical exaggeration to terrain in San Luis Obispo County, California.

Step 1. Start ArcScene and Add a Raster Layer

Double-click the ArcScene icon on your desktop, or open ArcScene from the Start menu.

Click the Add Data button. 

Navigate to your 3DDATA\Chapter02\Data folder. Double-click slo_cnty.lyr to add it to the scene. (For version 9.1 or 9.2, add slo_cnty_v91.lyr instead.)

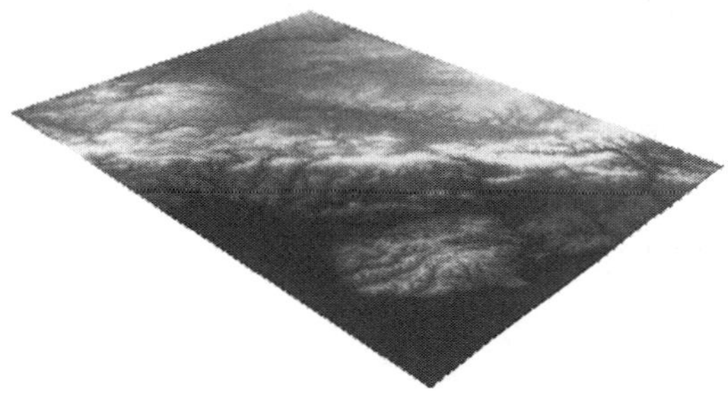

Step 2. Look at the Layer File and Its Data Source

Notice that the layer file's name in ArcScene's Table of Contents is slo_cnty, not slo_cnty.lyr, even though in the Add Data dialog box it is called slo_cnty.lyr. That's because by default, when you add a layer to a map, it is named after its data source. In this case, the data source is a Digital Elevation Model called slo_cnty.

Click the Add Data button again, and select the Details button on the Add Data toolbar. 

This view of the data tells you that slo_cnty is a raster dataset, and slo_cnty.lyr is a layer file.

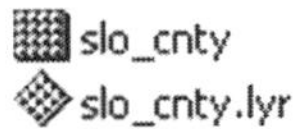

Raster Dataset
Layer

Now select the Thumbnails button. 

Scroll down in the Add Data window until you can see slo_cnty and slo_cnty.lyr.

Slo_cnty is the raw raster dataset. Slo_cnty.lyr is a file containing a set of instructions telling ArcScene, ArcMap, or ArcCatalog to draw slo_cnty with a certain color scheme, shading, and base height setting. When you add slo_cnty.lyr to ArcScene, it references slo_cnty. For this reason, if you were to move slo_cnty to another folder, slo_cnty.lyr would be unable to reference it. (If this happens, ArcScene gives you the option to "Set Data Source" by right-clicking the layer file in the Table of Contents. For more information on repairing data sources, search the ArcGIS Desktop Help for "Repairing the data source in a layer.")

Close the Add Data dialog without adding any data.

Step 3. Navigate the Scene

Click the Navigate tool. 

Use the left and right mouse buttons to navigate and zoom to the surface of slo_cnty.lyr. To pan, use both buttons together, or hold down the scroll wheel.

This area of San Luis Obispo is about 1200 square miles—an area so large that the 3160-foot difference between the highest and lowest elevation points looks small.

Step 4. Set a Vertical Exaggeration for the Scene

Click the Full Extent button. 

Move ArcScene to one side of your screen to make room for the Scene Properties dialog box.

Open the Scene Properties dialog. You can do this in any of three ways: by choosing Scene Properties from the View menu, by right-clicking "Scene layers" in the Table of Contents and choosing Scene Properties, or by double-clicking the words "Scene layers" in the Table of Contents.

Move the Scene Properties dialog so that you can see the layer in ArcScene.

Click the General tab.

In the Vertical Exaggeration box dropdown menu, choose 5.

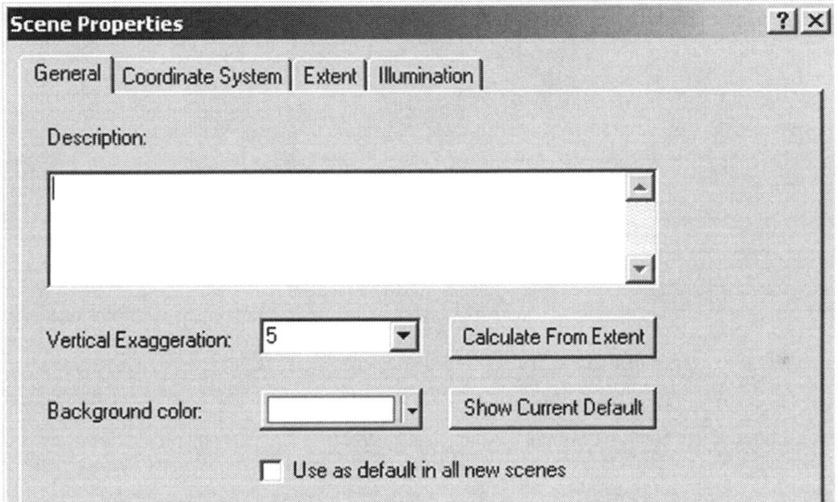

Click OK.

Use the Navigate tool to examine the surface. The hills are a little taller; the valleys are a little deeper.

Step 5. Have ArcScene Calculate Vertical Exaggeration

Open the Scene Properties dialog again, and make sure the General tab is active. Click Calculate From Extent.

ArcScene calculates a vertical exaggeration for you based on the extent of all the data displayed in the scene. In this case, it is about 8.34.

Click OK. (If the layer disappears from the scene, click the Full Extent button to bring it back into view.)

San Luis Obispo is beginning to look like the Alps. Notice that in the Table of Contents the elevation values have not changed.

As mentioned earlier, vertical exaggeration is a visual effect applied to all layers in a scene. It does not change data values or influence analysis.

Step 6. Change Vertical Exaggeration and Color Scheme

Open the Scene Properties dialog again. This time, change the vertical exaggeration to 25.

Click OK.

In ArcScene, click the Full Extent button to center the layer in the display.

In the Table of Contents, click once on the color ramp for slo_cnty. This brings up the Select Color Ramp dialog.

Right-click the color ramp in the dialog box, and uncheck "Graphic View." This gives you the names of the color ramps.

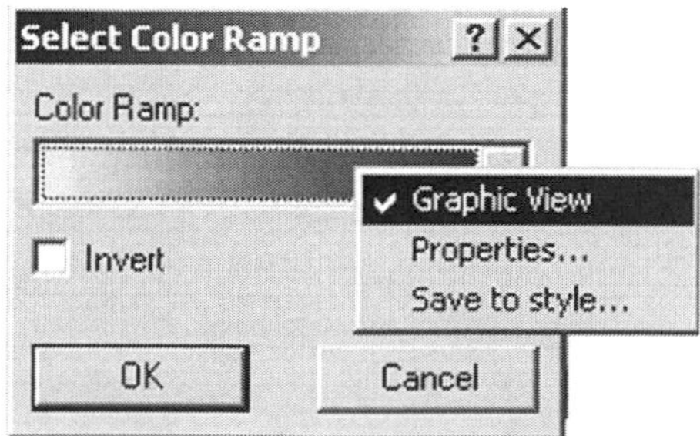

Scroll down through the ramp descriptions and choose "Spectrum-Full Bright."

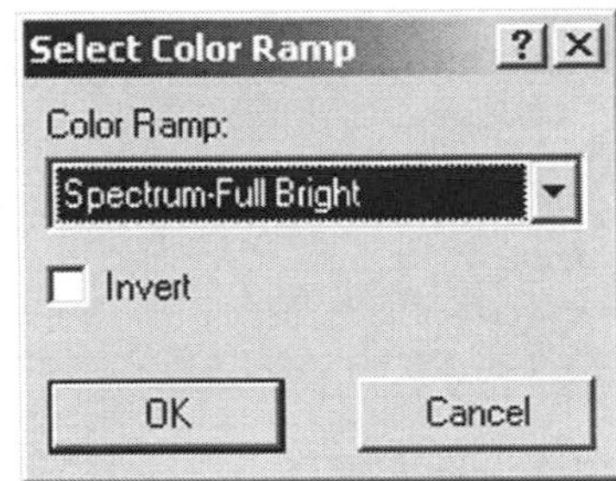

Click OK in the Select Color Ramp dialog.

Step 7. Navigate Again

Use the Navigate tool to take a look at your crazy new landscape. If you always wanted a set of Magic Rocks as a kid, now you can use ArcScene to grow your own.

Step 8. Close ArcScene

When you're finished navigating, close ArcScene. Click No when you're asked if you want to save changes.

Exercise 2-3

Apply a Coordinate System to a Scene

When you add a layer to an empty scene, the scene adopts that layer's coordinate system. If you add subsequent layers that use different coordinate systems, ArcScene temporarily projects the layers to fit the scene. As in ArcMap, this does not affect the coordinate systems of those layers' source data.

In this exercise, you will add a TIN of an area near Coletown, KY, to a scene. You'll change the scene's coordinate system from feet to meters, and change the vertical exaggeration to compensate for the visual impression created by displaying z values in meters instead of feet.

Step 1. Start ArcScene

Double-click the ArcScene icon on your desktop, or open ArcScene from the Start menu.

Click the Add Data button.

Navigate to your 3DDATA\Chapter02\Data folder. Double-click cole_tin.lyr to add it to the scene. (If you're using version 9.1 or 9.2, add cole_tin_v91.lyr instead.)

Step 2. Look at Cole_tin's Coordinate System in Layer Properties

You can find a layer's coordinate system information in several different places. Right-click cole_tin in the Table of Contents and select Properties. In the Layer Properties dialog, click the Source tab.

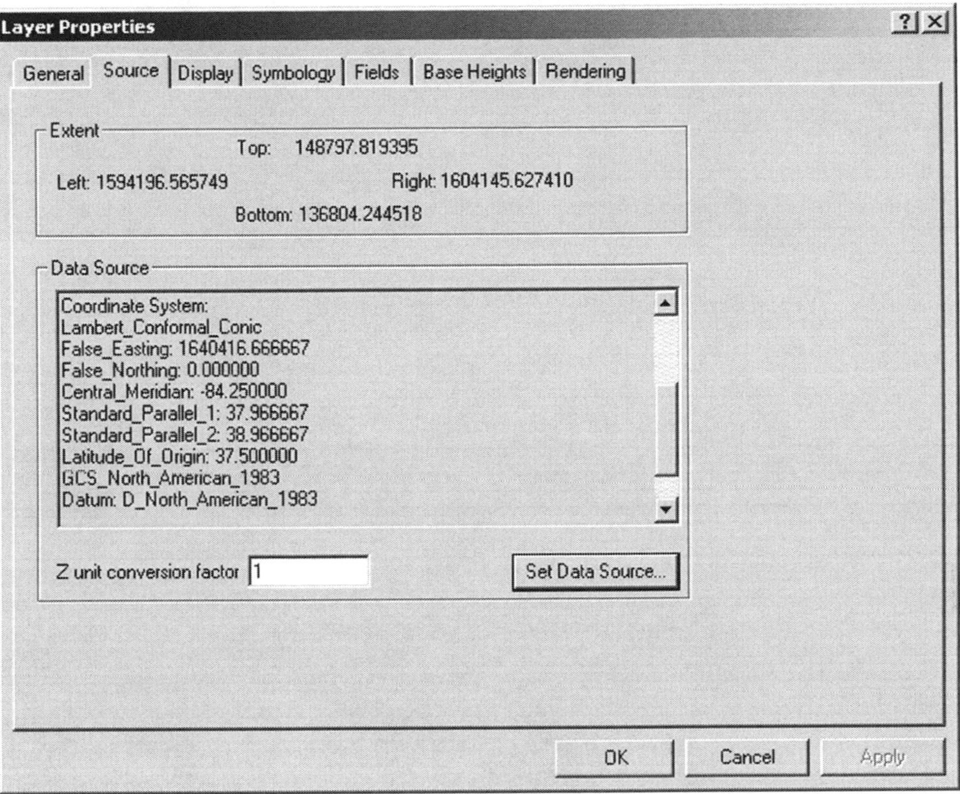

Cole_tin uses a Lambert Conformal Conic projection based on the NAD 1983 datum.

Click OK to dismiss the Layer Properties dialog.

Step 3. Look at Cole_tin's Coordinate System in ArcCatalog

In ArcScene, click the ArcCatalog button. 

In ArcCatalog, navigate to your 3DDATA\Chapter02\Data folder. Select cole_tin.

Click the Metadata tab. Within the Metadata window, click the Spatial tab.

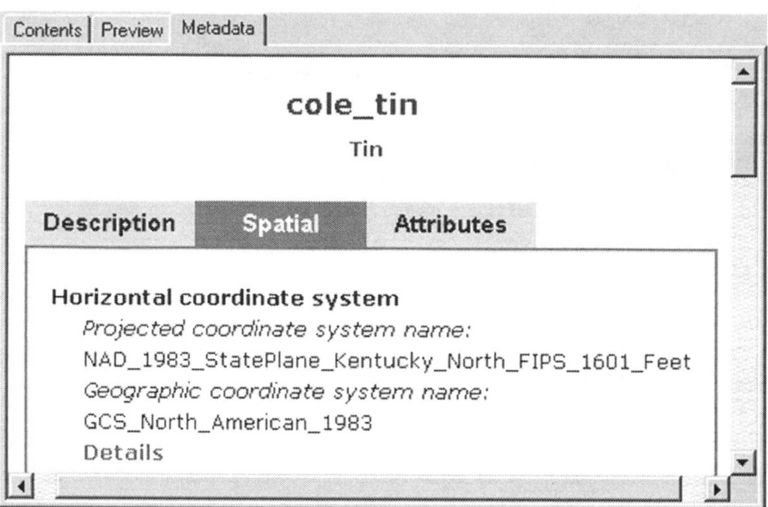

Cole_tin is in the NAD 1983 State Plane coordinate system.

Click the Details button to see all of the coordinate system metadata.

The State Plane coordinate system is designed for large-scale (small area) U.S. mapping that divides the 50 states into over 120 numbered zones, each with its own projection parameters. The Lambert Conformal Conic projection is used for states that are longer in the east-west direction, like Kentucky and Tennessee, and the Transverse Mercator projection is used for states that are longer in the north-south direction like Vermont and Illinois. Notice that the units are survey feet.

Close ArcCatalog.

Step 4. Look at Cole_tin's Coordinate System in Scene Properties

In the Table of Contents, right click Scene Layers and choose Scene Properties.

Click the Coordinate System tab.

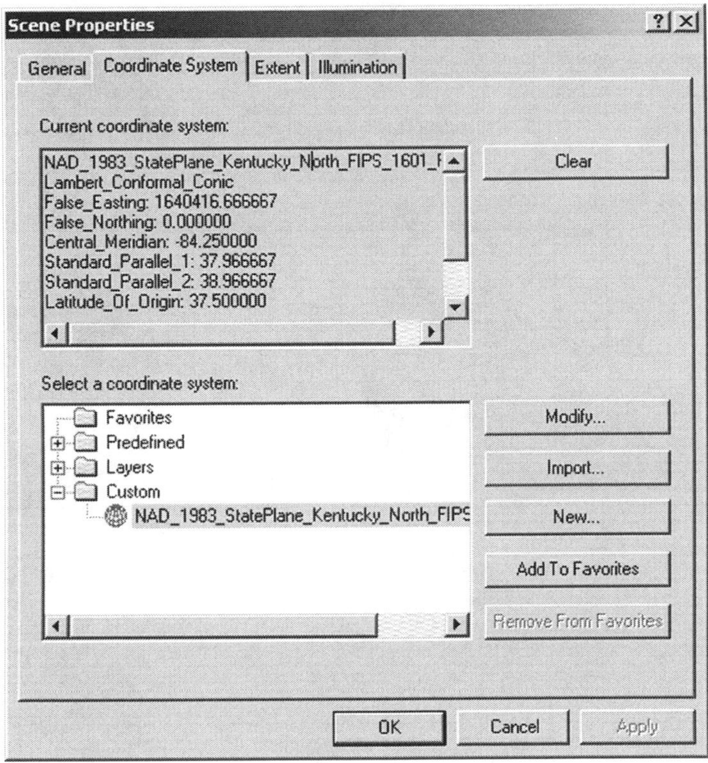

Because cole_tin is the first layer added to the scene, you know the scene is using cole_tin's coordinate system. When you change the scene's coordinate system in a few minutes, you'll see that the scene no longer reflects cole_tin's coordinate system.

Close the Scene Properties dialog.

Step 5. Zoom in to the Kentucky River

Use the Navigate tool to zoom in, pan, and get a good look at the banks of this particular bend in the Kentucky River.

Click the Identify tool. 

Click on a few places along the river. Notice that it is about 540 feet above sea level. The x and y coordinates are also in feet—take a look at the range of numbers you get with the Identify tool at various locations.

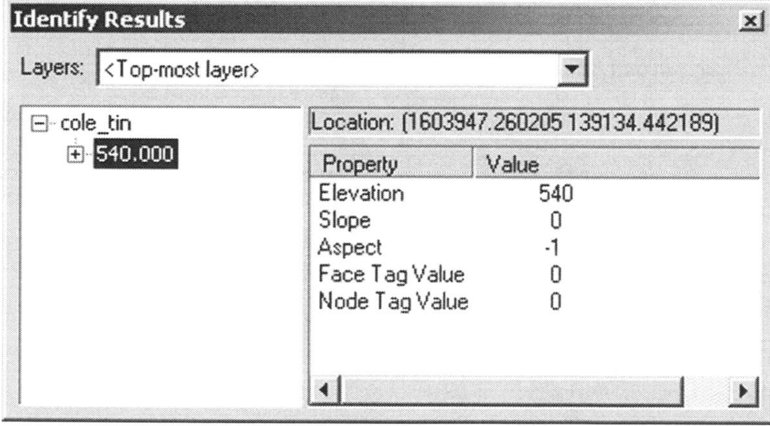

When you're finished examining the surface with the Identify tool, close the Identify Results box.

Step 6. Change the Scene's Projection

Now you are going to project the scene into the UTM zone that is appropriate for the location of cole_tin.

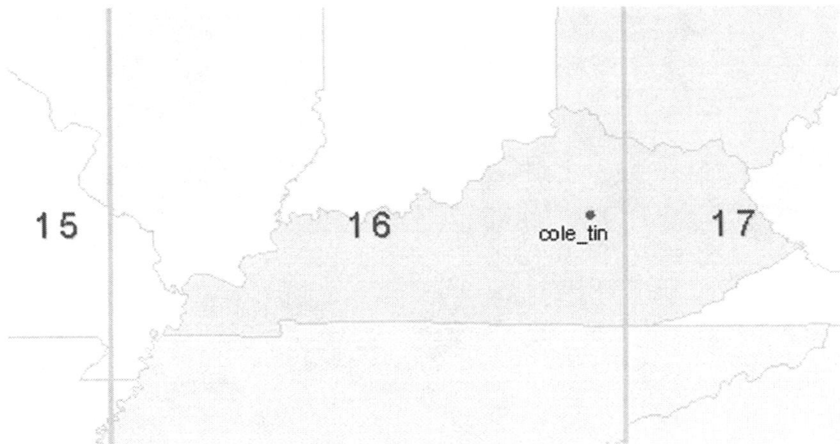

The Universal Transverse Mercator system divides the globe into 60 zones running north-south, each spanning six degrees of longitude. Each zone has its own central meridian and is split by the equator into north and south sections. As you can see from the map above, cole_tin lies in zone 16.

Open the Scene Properties dialog (right-click or double-click Scene Layers in the Table of Contents, or choose Scene Properties from the View menu).

The Coordinate System tab should still be selected. In the window under "Select a coordinate system" click on "Predefined," then click on "Projected Coordinate Systems," then "UTM," then "NAD 1983."

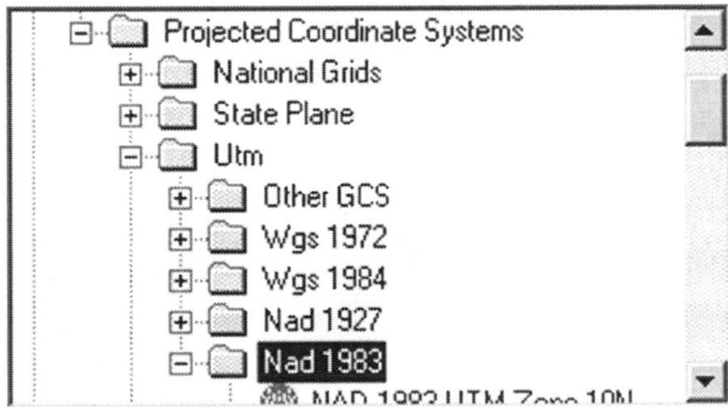

You're choosing UTM NAD 1983 because the Kentucky State Plane coordinate system that cole_tin uses is also based on the NAD 83 datum.

Scroll down in the list of NAD 1983 UTM zones and select Zone 16N. (The N is for North, since Kentucky is north of the equator.)

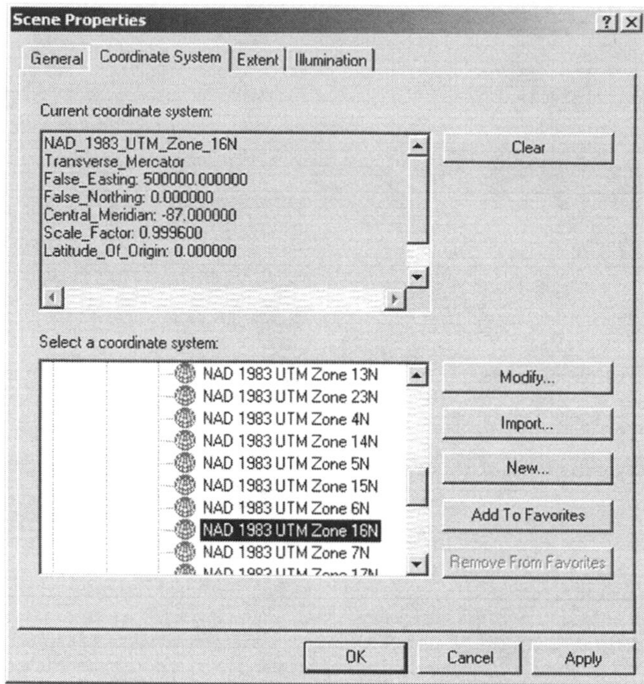

Click OK to change the scene's coordinate system from Kentucky State Plane to UTM Zone 16N.

Step 7. Examine the Scene with a UTM Projection

Use the Navigate tool to look at cole_tin again. It looks a lot lumpier. Zoom to the bend in the river. The banks have become a canyon. What's going on?

The explanation can be found by looking at the properties of the new UTM projection.

Double-click Scene Layers to bring up Scene Properties again. The Coordinate System tab should be selected.

NAD_1983_UTM_Zone_16N is the scene's current projection. You may recall that when the scene was in Kentucky State Plane, the units—feet—were part of the name of the coordinate system and so showed up in the title under "Current Coordinate System." You can't tell from the title here what units the coordinate system is in, though, so you'll have to dig a little deeper.

With NAD_1983_UTM_Zone_16N selected, click Modify.

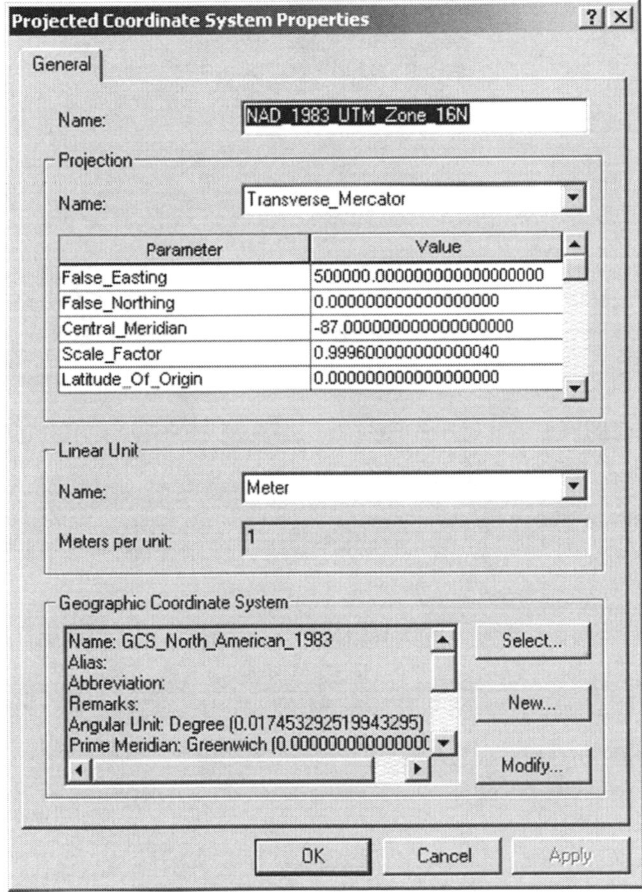

Under "Linear Unit" notice that the units for this projection are meters, not feet.

Click Cancel to close the Projected Coordinate System Properties dialog without making any modifications. (If you accidentally click OK, don't worry about it—you'll just have a duplicate listing of the current coordinate system.)

Click OK to close the Scene Properties dialog.

Step 8. Identify Locations in the River Again

When you changed the scene's coordinate system from State Plane to UTM, the x,y-units were successfully changed from feet to meters, so horizontal measurements taken in either projection are properly converted. Changing the coordinate system doesn't change the z-units, however.

Click the Identify tool. Click on some locations in the riverbed again.

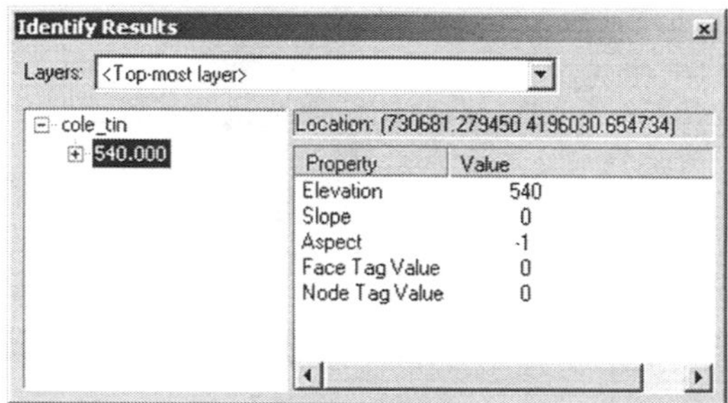

As you saw earlier, the elevation of the Kentucky River at this point is about 540 feet. If the z-units had been converted to meters, they would report the height of the river at about 180, or about third of 540. But the Identify results still report a value of 540. We know that they're feet, but ArcScene thinks they're meters, so the z-units of cole_tin look three times taller than they should.

Step 9. Compensate for the Feet-to-meters Discrepancy by Changing the Scene's Vertical Exaggeration

There are several ways to fix the way cole_tin appears in the scene. One way is to go back to the Modify button in the Projected Coordinate System Properties dialog and change "Meter" to "Foot." Another way is to change the vertical exaggeration of the scene.

Double-click Scene Layers in the Table of Contents to bring up the Scene Properties dialog.

Click the General tab.

Highlight "None" in the Vertical Exaggeration box, and type in the value 0.3048. This effectively converts the z-units in the scene from feet to meters.

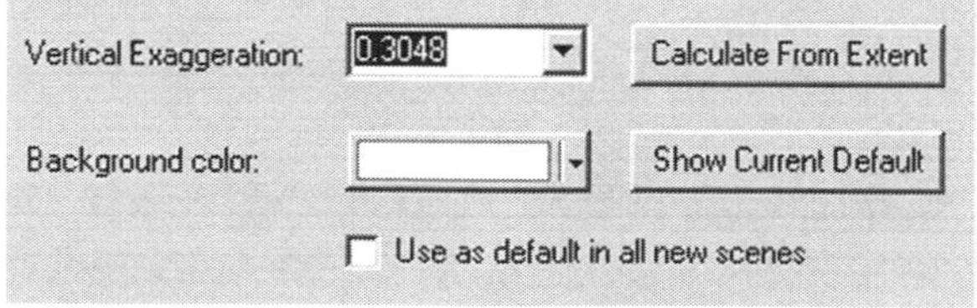

Click OK in the Scene Properties dialog.

Voila! Cole_tin looks normal again.

Vertical exaggeration is just a visual effect, remember. If you use the Identify tool, the z measurements will be reported in feet, and the x,y measurements will be reported in meters. That's because ArcScene gets the x,y units from the scene's projection but gets the z units from the coordinate system of cole_tin's source data, which as you know is State Plane Feet. To see the elevation reported in meters, you'd need to alter the original cole_tin dataset.

Step 10. Close ArcScene

Close the Identify Results box and close ArcScene. Say No when asked if you want to save your changes.

Exercise 2-4

Set 3D Layer Properties for an Elevation Raster

As you learned in the beginning of Chapter 2, you have to set the base heights of a raster layer when you add it to a scene. In this exercise, you'll add a raster to ArcScene and set its base heights, shading, and symbology.

Step 1. Open ArcScene

Double-click the ArcScene icon on the desktop, or open ArcScene from the Start menu.

Click the Add Data button in ArcScene. 

Navigate to your 3DDATA\Chapter02\Data folder. Double-click cole_ras to add it to the scene.

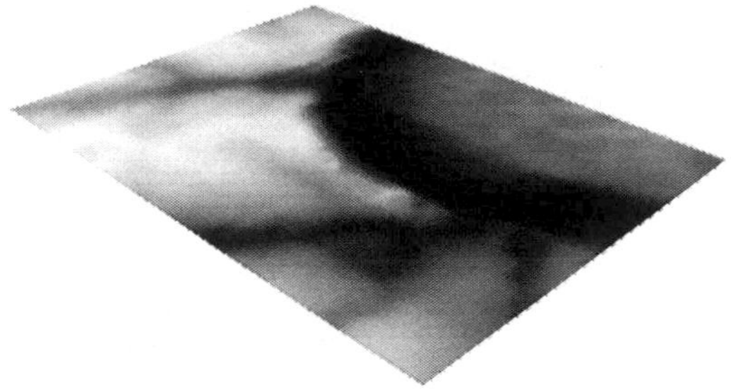

Step 2. Examine the Data

In the Table of Contents, right-click cole_ras and choose Open Attribute Table.

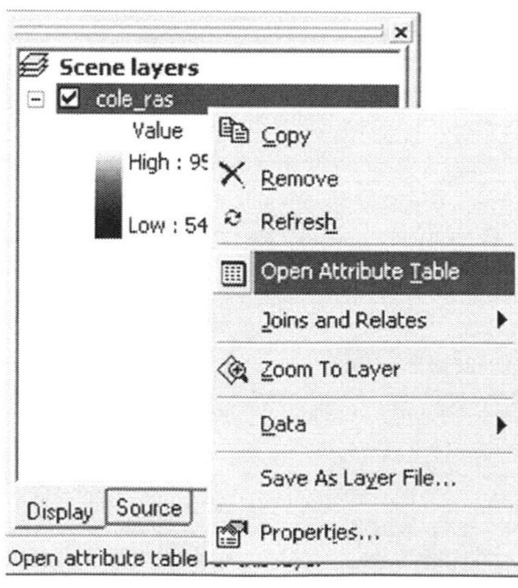

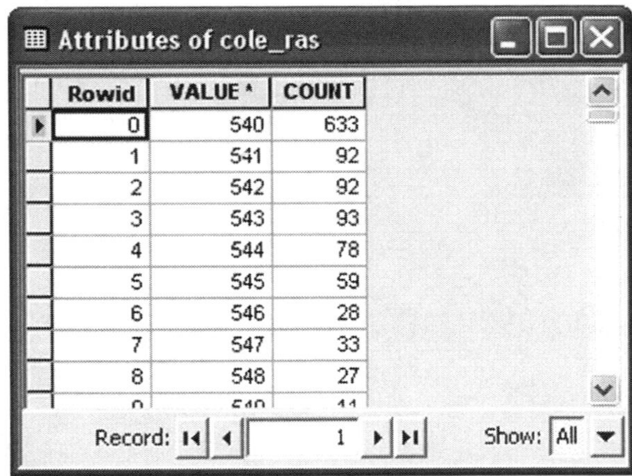

Rowid	VALUE *	COUNT
0	540	633
1	541	92
2	542	92
3	543	93
4	544	78
5	545	59
6	546	28
7	547	33
8	548	27

Cole_ras is a small portion of a DEM of Coletown, KY. You worked with a TIN made from that DEM in the last exercise. Cole_ras has 414 records, one for each unique elevation value in the raster. The elevations are in State Plane Feet and are stored in the Value field. For each record, the Count field shows how many cells have that elevation.

Close the table.

Step 3. Set Base Heights

When you add a raster to ArcScene, you will generally have to set its base heights, shade it, and symbolize it. You do all of these through the Layer Properties dialog.

Move ArcScene to the right side of your screen so that you can still see cole_ras when you open the Layer Properties dialog box.

Right-click cole_ras in the Table of Contents and choose Properties. Move the dialog to the left of your screen.

Click the Base Heights tab.

As you may remember, base heights for rasters are set to 0 by default, so the layer starts out flat.

Choose "Obtain heights for layer from surface," and pick cole_ras from the dropdown list (it should be the only dataset listed).

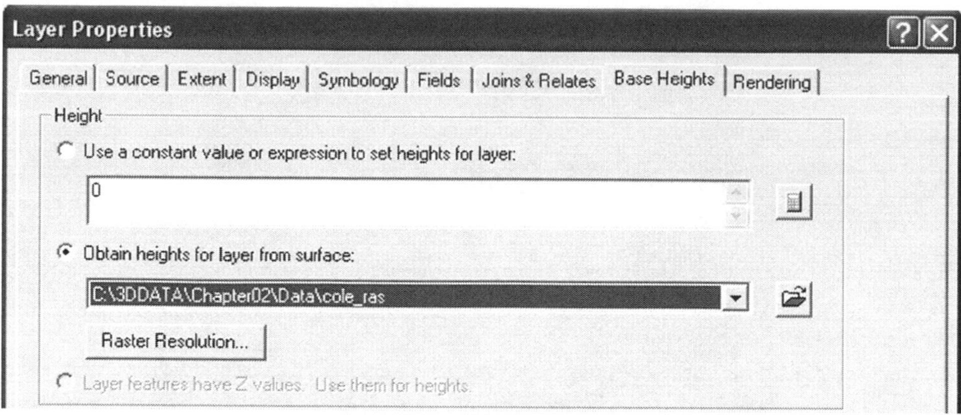

Click Apply, and leave the dialog box open.

Now ArcScene displays the layer according to the elevation values you just examined in the Value field of the cole_ras attribute table.

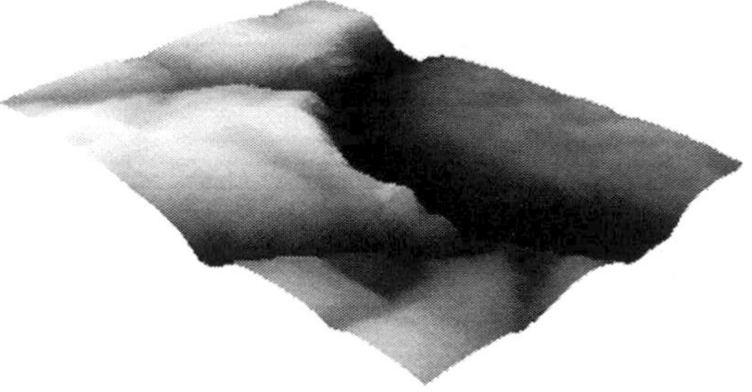

Step 4. Set the Shading for Cole_ras

In the Layer Properties dialog, click the Rendering tab.

In the Effects panel, check the box next to "Shade areal features relative to the scene's light position." "Use smooth shading if possible" is checked automatically.

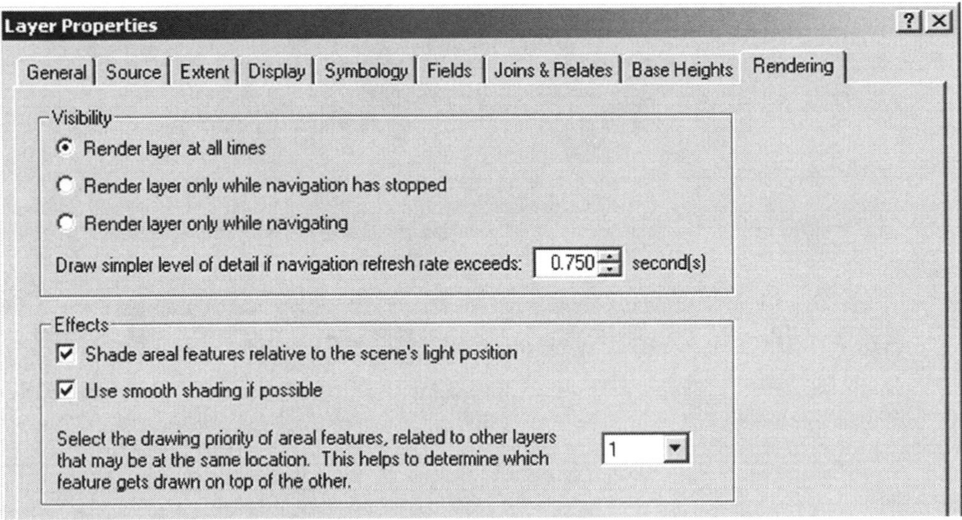

Click Apply again.

Now the raster is shaded from the northwest. It looks a bit dark, so in a minute you'll raise the sun to a higher altitude.

Step 5. Change the Symbology of Cole_ras

Click the Symbology tab on the Layer Properties dialog.

Right-click the color ramp so that you can uncheck the "Graphic view" box.

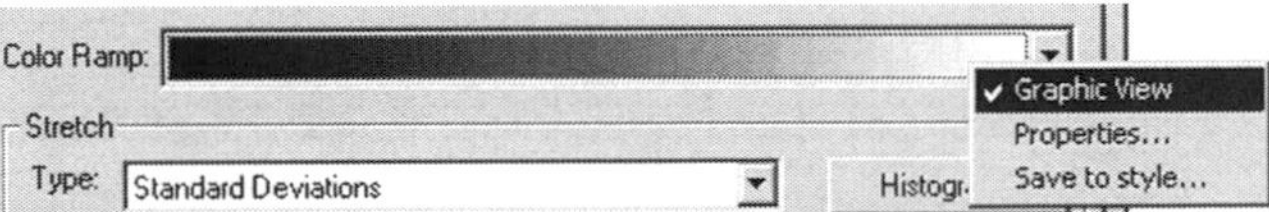

Scroll through the list of color ramps and choose "Yellow to Green to Dark Blue."

Under Stretch, change the "n" value to "1." Increasing the value of *n* shades more cells with colors from the middle of the ramp, while decreasing the value of *n* shades more cells with colors at each end of the ramp.

Click OK. This closes the Layer Properties dialog box. Now cole_ras is displayed in 3D, shaded, and symbolized according to its elevation values.

Step 6. Add a Background Color and Change the Sun's Altitude

In the Table of Contents open the Scene Properties, either by double-clicking Scene layers, or by right-clicking Scene layers and choosing Scene Properties.

Click the General tab.

Click the arrow next to the Background color box, and choose a color you like—perhaps a nice blue.

Click Apply, and leave the Scene Properties dialog open.

Now, click the Illumination tab.

In the Altitude box, highlight the value and type in "40."

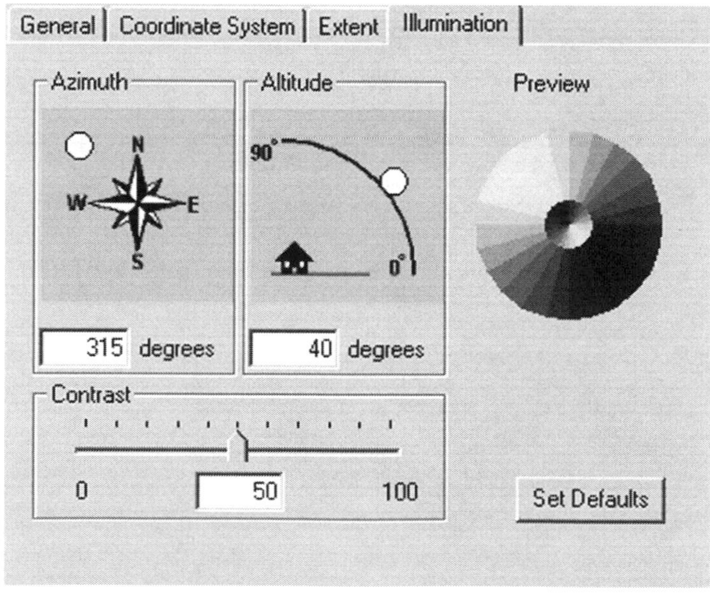

Click OK to close the Scene Properties dialog.

Step 7. Navigate the Dataset

Cole_ras is more brightly illuminated for viewing. Click the Navigate tool and zoom in and around the raster.

Step 8. Save the ArcScene Document

From the File menu, choose Save. Navigate to your 3DDATA\Chapter02\MyData folder. Give the ArcScene document a name of your choice, and click Save. The ArcScene document is saved in your Chapter02\MyData folder.

Close ArcScene.

Exercise 2-5

Set 3D Layer Properties for a Raster Image

In the last exercise, you added an elevation raster to ArcScene and set its base heights so that it would display in 3D according to its own elevation values. In this exercise, you'll drape an image raster over the elevation raster.

Step 1. Open an ArcScene Document

Start ArcScene. From the File menu choose Open, and navigate to your 3DDATA\Chapter02\Data folder.

Double-click raster_heights.sxd. This is a copy of the ArcScene document that you made in the last exercise. (For version 9.1 or 9.2, open raster_heights_v91.sxd instead.)

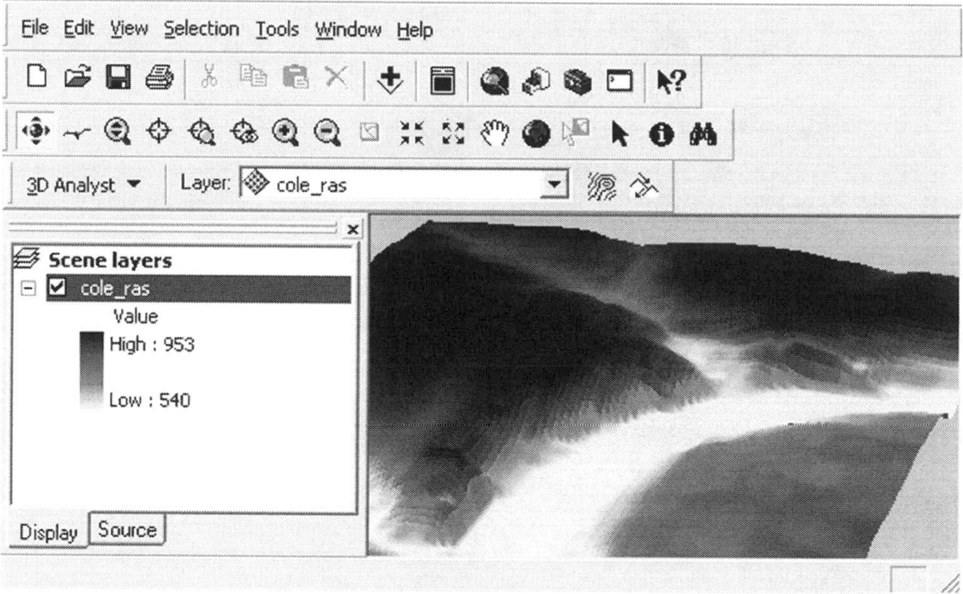

Step 2. Add an Aerial Photograph

Click the Add Data button. You should still be in your 3DDATA\Chapter02\Data folder.

Select cole_doq.jpg, and add it to the scene.

Click the Full Extent button. 

The new raster's base heights are set to 0 by default when it is first added to ArcScene, so it looks flat.

Cole_doq is a Digital Orthophotoquad, an aerial photograph that has been geometrically rectified to account for distortions owing to the camera's tilt and uneven terrain. This means that distances and directions between landmarks in the photo are proportional to distances and directions measured on the ground.

This doesn't mean that cole_doq.jpg has elevation values in its cells, however. It's an image, so its cells contain light information. In ArcScene, you can borrow the base heights from an elevation raster or a TIN and use them as the base heights for an image, for 2D vector data, or for other rasters that represent phenomena besides elevation. The elevation surface is typically present in the scene, but it can just be a layer that you browse to on disk.

Step 3. Set Base Heights for the Aerial Photo

In the Table of Contents, right-click cole_doq.jpg and choose Properties.

Click the Base Heights tab. Click "Obtain heights for layer from surface," and in the dropdown list, choose cole_ras (not cole_doq.jpg).

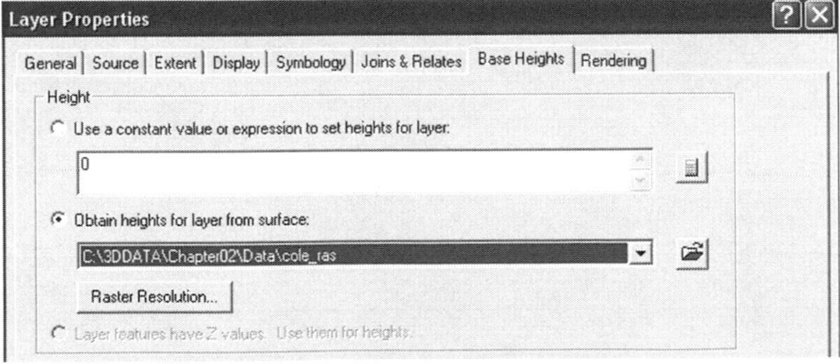

Click OK to dismiss the Layer Properties dialog box.

The aerial photo is draped over the elevation raster. It looks strange because the base heights of the photo are set to the base heights of the elevation raster, so they're competing for the same display space.

There are a couple of ways to deal with this. The simplest is to turn off the elevation layer.

In the Table of Contents, uncheck the box next to cole_ras.

Now you can see the aerial photo by itself.

Step 4. Add a Height Offset to the Photo

Suppose that you want to see the photo on top, but you'd still like to be able to look at the elevation raster. You can see both by adding an offset to the image layer.

Turn cole_ras on again in the Table of Contents.

Double-click cole_doq.jpg to bring up its layer properties. The Base Heights tab should be selected.

In the Offset panel, type "1000." This raises the image layer 1000 units—in this case, feet—above the elevation raster.

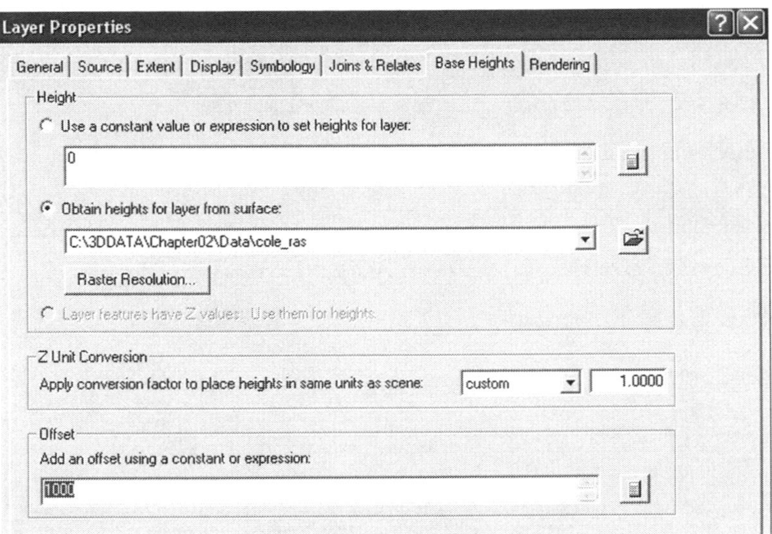

Click OK.

Now you can easily examine both surfaces. Use the Navigate tool to pan and zoom around the layers.

Step 5. Save the ArcScene Document

From the File menu, choose Save As. Give the document a new name and save it in your 3DDATA\Chapter02\MyData folder. Then close ArcScene.

Exercise 2-6

Set Base Heights for a 2D Vector Layer

In the last exercise, you used an elevation raster to set base heights for an image. In this exercise, you'll use an elevation raster to set the base heights for 2D vector features.

Step 1. Start ArcScene and Add an Elevation Raster

Double-click the ArcScene icon on your desktop, or open ArcScene from the Start menu.

Click the Add Data button, and navigate to your 3DDATA\Chapter02\Data folder.

Double-click harlan_dem to add it to the scene.

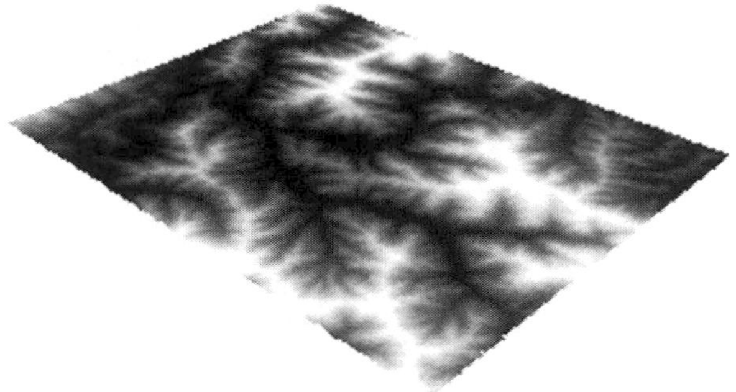

You worked with this DEM of Harlan County, KY, in Chapter 1.

Step 2. Add a 2D Shapefile

Click the Add Data button again. Double-click harlan_creeks.shp to add it to the scene.

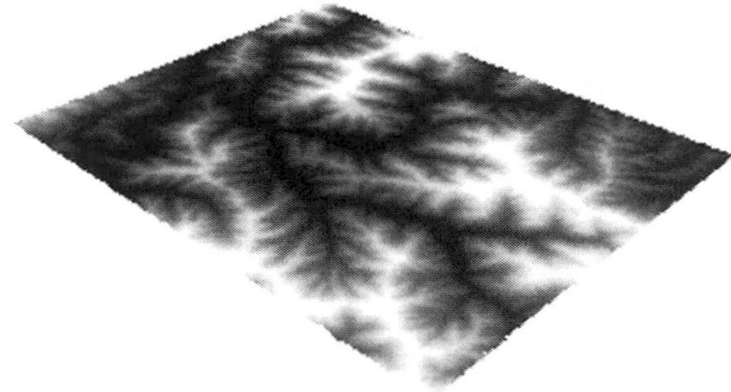

The raster's base heights are set to 0, and the creeks are 2D polyline features, so both layers lie flat.

Step 3. Change the Line Symbol for Harlan_creeks

In the Table of Contents, click on the line symbol underneath harlan_creeks.

This brings up the Symbol Selector dialog. Find the "River" line symbol, and select it.

Click OK. Now the creeks show up more clearly against the DEM.

Step 4. Examine the Attribute Table of the Harlan_creeks Shapefile

In the Table of Contents, right-click harlan_creeks and choose Open Attribute Table.

Attributes of harlan_creeks

FID	Shape*	NAME	LENGTH
0	Polyline	MARTINS FORK CUMBERLAND RIVER	13.62013
1	Polyline	MARTINS FORK CUMBERLAND RIVER	13.62013
2	Polyline	BEE BRANCH	0.70678
3	Polyline	CATRON CREEK	6.29786
4	Polyline	BAKER BRANCH	0.82363
5	Polyline	POUNDING MILL BRANCH	1.06721
6	Polyline	BOBS CREEK	1.96592
7	Polyline	JERRY BRANCH	0.98673
8	Polyline	IRVIN BRANCH	0.9357

Record: 0 Show: All Selected Records (0 out of 69 Selected.)

The shapefile has 69 records. You can tell that it's a 2D shapefile because the records in the Shape field say Polyline. If it were a 3D shapefile, they would say PolylineZ (or PolylineZM).

Close the attribute table.

Step 5. Set the Raster's Display Properties

In the Table of Contents, right-click harlan_dem and choose Properties.

Click the Base Heights tab.

Check "Obtain heights for layer from surface" and choose harlan_dem as the source in the dropdown list.

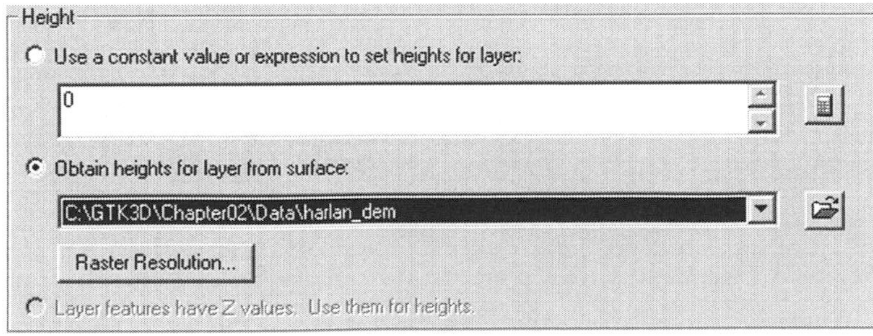

Click the Rendering tab.

In the Effects box, choose "Shade areal features relative to the scene's light position."

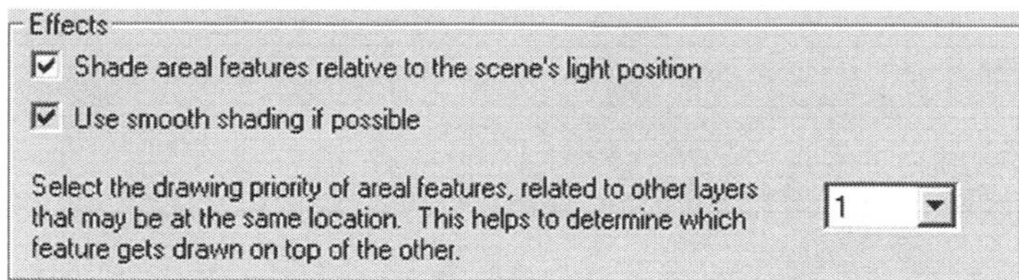

Finally, click the Symbology tab, and right-click the color ramp so you can uncheck "Graphic View."

Scroll through the color ramps descriptions, and choose "Green Light to Dark." Check the box next to "Invert" so that the darker shades will represent the lower elevations.

Click OK to activate the settings, and close the Layer Properties dialog.

Harlan_dem is displayed in 3D, but the creeks can't be seen.

Step 6. Set Base Heights for the Creeks Layer

Use the Navigate tool to zoom into the scene, and take a look underneath harlan_dem. If you don't see the underside of the raster, it's because the "Layer Face Culling" setting is hiding the back of the layer. To fix this, go to the View menu in ArcScene, then choose Toolbars, and make sure that the 3D Effects toolbar is checked.

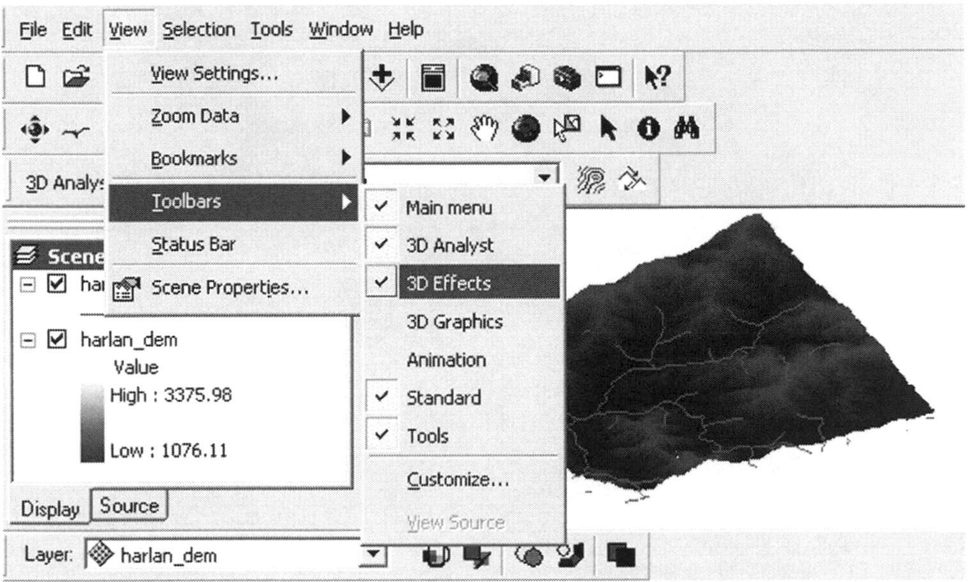

On the 3D Effects toolbar, make sure Harlan_dem is selected in the dropdown list, and click the Layer Face Culling button. Choose "View both sides of areal features."

The creeks haven't had their base heights set yet, so they're hidden under harlan_dem.

In the Table of Contents, right-click harlan_creeks and choose Properties.

Click the Base Heights tab.

Click "Obtain heights for layer from surface," and select harlan_dem for the source of heights in the dropdown menu (it should be your only choice).

Click OK, and click the Full Extent button.

Now the 2D creeks are using harlan_dem's elevation values as base heights, so you can see them on top. Each vertex in the polyline shapefile adopts the elevation value from the corresponding cell in harlan_dem.

Step 7. Navigate the Scene

Zoom in and look at the places where the creeks flow through the hills. In certain areas and from certain angles, the creeks disappear into the terrain, giving it a stitched effect.

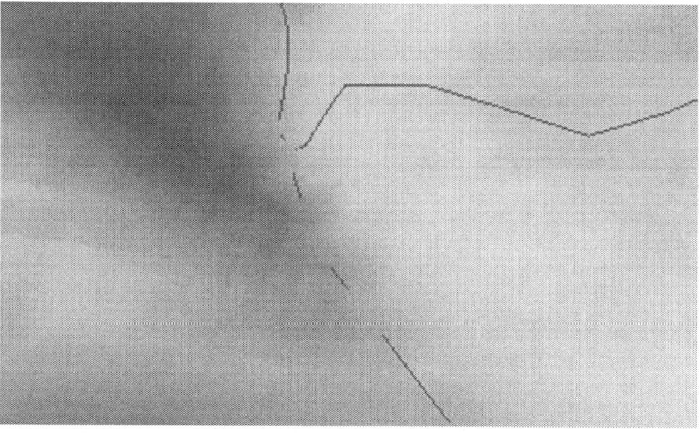

If you turn off harlan_dem, you can see that the creeks are continuous, but that the lines between the vertices are being buried by the variations in the raster layer.

Step 8. Change Harlan_dem's Drawing Priority

In the last exercise, you saw how to offset layers that are competing for the same 3D space. You can also set a layer's drawing priority so that it displays above or below other layers.

Move ArcScene to the right side of your screen.

In the Table of Contents, right-click harlan_dem and choose Properties.

Click the Rendering tab, and move the Layer Properties dialog to the left, out of the way of the scene.

In the Effects frame of the dialog, change the layer's drawing priority to 8. A value of 1 is the highest priority and is the default for all layers. 10 is the lowest drawing priority setting. By setting harlan_dem to 8, you tell it to draw underneath harlan_creeks.shp.

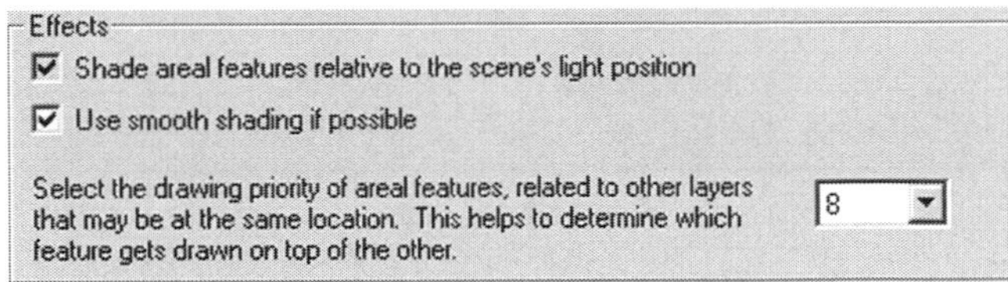

Click OK.

The elevation raster now has the lowest drawing priority in the scene. More of the creeks appear.

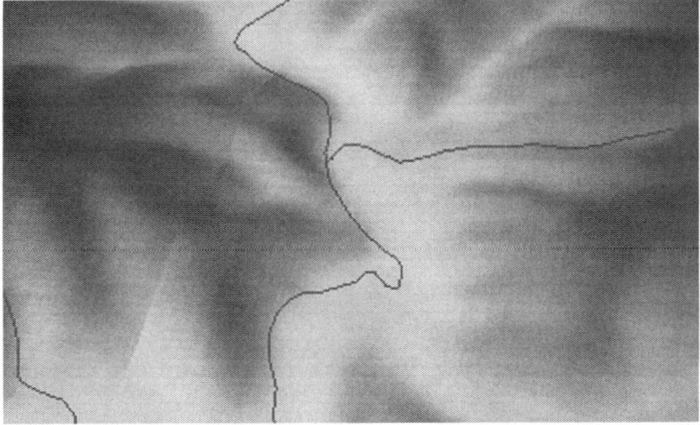

Step 9. Offset the Shapefile

In the Table of Contents, right-click harlan_creeks and choose Properties.

Click the Base Heights tab, and move the Layer Properties dialog to the left of the scene.

In the Offset panel, type "25" into the box. This raises the creek layer 25 feet above the raster layer.

Offset
Add an offset using a constant or expression:
25

Click OK.

Now the stitching is gone, and the creeks are displayed fully on top of the raster.

Step 10. Navigate again

Click the Full Extent button.

Use the Navigate button to look at the creeks from various angles. You should be able to see them on top of the raster layer.

In the Table of Contents, right-click harlan_dem and choose Refresh.

(If the raster appears "holey" you can set its drawing priority to a higher number, and give the creeks more of an offset to compensate. As a hint, the first operation is done the Layer Properties dialog for harlan_dem; the second is done in the Layer Properties dialog for harlan_creeks.)

Step 11. Save the ArcScene Document

From the File menu, choose Save As.

Give the document a name of your choice and save it in your 3DDATA\Chapter02\MyData folder.

Close ArcScene.

Exercise 2-7

Extrude 2D Vector Features

In the last exercise, you used an elevation raster to set the base heights for 2D line features. In this exercise, you'll use an elevation raster to set the base heights for 2D polygons, and you'll extrude them by depth values provided in their attribute table. You may remember from the discussion earlier in Chapter 2 that when vector features are extruded, points become vertical lines, lines become walls, and polygons become solids.

Step 1. Start ArcScene and Add an Elevation Raster

Double-click the ArcScene icon on your desktop, or open it from the Start menu.

Click the Add Data button, navigate to your 3DDATA\Chapter02\Data folder. Double-click slo_cnty to add it to the scene.

This is the elevation raster you used earlier, covering an area of about 1200 square miles in San Luis Obispo County, California.

Step 2. Set Base Heights for the Raster Layer

Right-click slo_cnty in the Table of Contents and choose Properties.

Click the Base Heights tab. Click "Obtain heights for layer from surface," and select slo_cnty as the source for heights.

Click OK.

Step 3. Set Vertical Exaggeration and Background Color for the Scene

In the Table of Contents, double-click Scene Layers to bring up the Scene Properties.

Click the General tab. In the Vertical Exaggeration box, set the exaggeration to 7.

In the Background color box, choose Apatite Blue.

Click OK.

Step 4. Navigate the Layer

Click the Navigate tool, zoom in, and take a look around the layer.

Step 5. Add the 2D Polygons

Click the Add Data button again. You should still be in the 3DDATA\Chapter02\Data folder. Double-click mines.shp to add it to the scene.

Click the Full Extent button.

You can't see the mines because their base heights haven't been set, and they're hiding under slo_cnty.

Step 6. Change the Color of the Mine Polygons

In the Table of Contents, turn off slo_cnty. Right-click the polygon symbol under mines to bring up the color palette.

Choose Mars Red.

Step 7. Examine the Mines

Now you should be able to see the mines. Use the Zoom In tool to draw a rectangle around a few clustered together.

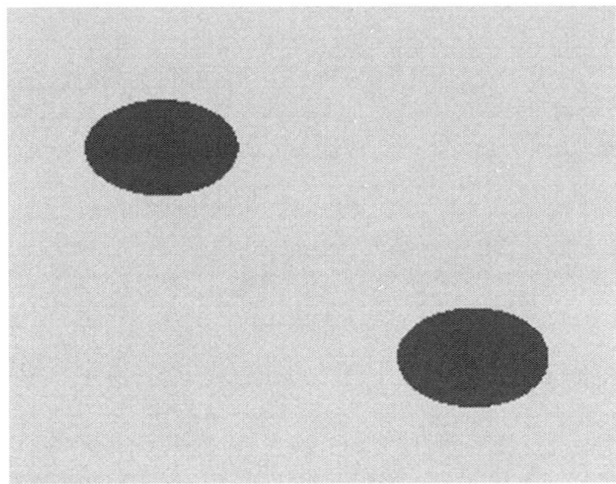

Use the Navigate tool to examine them more closely. Remember that by holding both mouse buttons down together, you can pan the scene.

These polygons are circles that were created by buffering a point shapefile of mine locations in San Luis Obispo. Each has a diameter of about 2350 feet, or almost half a mile. They don't reflect the actual acreage of the mines any more than the points did, but as polygons they are easier to see from a distance.

Step 8. Look at the Mine Attributes

In the Table of Contents, right-click mines and choose Open Attribute Table.

Attributes of mines

FID	Shape*	MINE1	MINETYPE	N_ELEV
12	Polygon	WHALE ROCK PIT	OPEN PIT	300
6	Polygon	TIBER CANYON SAND PIT	OPEN PIT	450
7	Polygon	TEMPLETON/ORMONDE	GRAVEL	740
14	Polygon	SYCAMORE ROAD PIT	GRAVEL	850
5	Polygon	SANTA RITA STONE QUARRY	QUARRY	140
16	Polygon	SANTA MARGARITA QUARRY	QUARRY	100
20	Polygon	ROCKY CANYON QUARRY	QUARRY	107
13	Polygon	PATCHETT PIT	OPEN PIT	540

Record: 0 Show: All Selected Records (0 out of 23 Selected.)

Mines.shp has 23 records. If you scroll to the right of the attribute table, you'll see a field called N_ELEV that contains the negative elevation, or lowest depth, of each mine.

Also look at the field called MINETYPE. There are three types of mines: quarry, open pit, and gravel.

Close the attribute table.

Step 9. Set the Base Heights for the Mines

Slo_cnty should still be turned off in the scene, and you should still be zoomed in closely to a few of the mine polygons.

In the Table of Contents, right-click mines and choose Properties.

Click the Base Heights tab, click "Obtain heights for layer from surface," and select slo_cnty as the source for heights.

Click OK.

The polygons are using the base heights from slo_cnty, so they're draped over the surface of the raster. Use the Navigate tool to look at how the shapes of the circles are changed by the terrain.

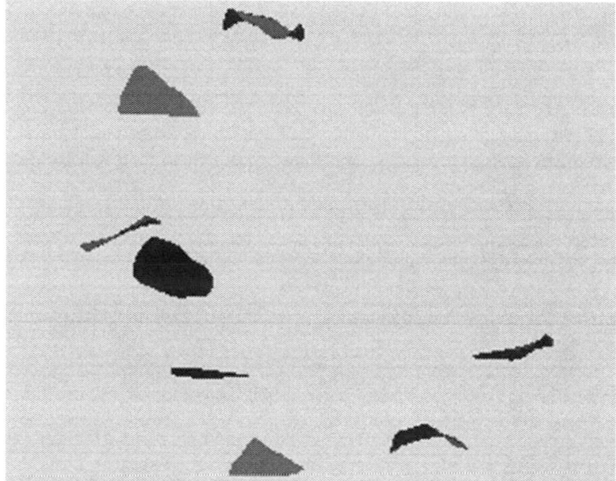

Step 10. Extrude the Polygons

Click the Full Extent button, and turn slo_cnty back on in the Table of Contents.

Right-click mines and choose Properties.

In the Layer Properties dialog, click the Extrusion tab.

When you looked at the attribute table for mines.shp, you saw the field called N_ELEV that contains the mine depths. You'll extrude the mines by their depth values.

Click the box next to "Extrude features in layer."

Click the Calculator button to open the Expression Builder dialog. 

In the Fields box, click N_ELEV to add it to the Expression box. Because you want to extrude the mines downward, you need to put a minus sign (a hyphen) in front of the expression. The extrusion value will be applied to each feature's minimum height.

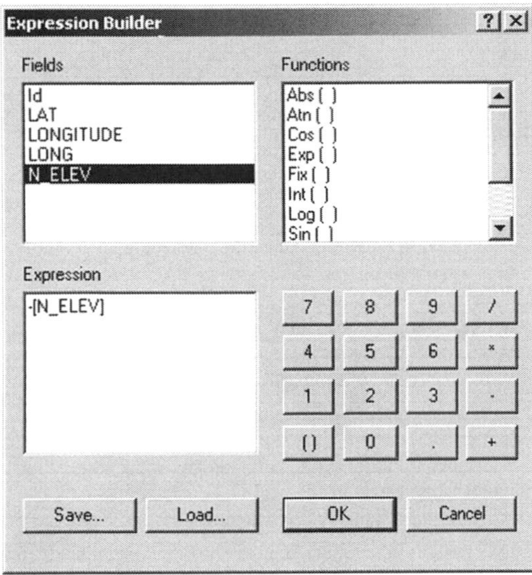

Click OK in the Expression Builder.

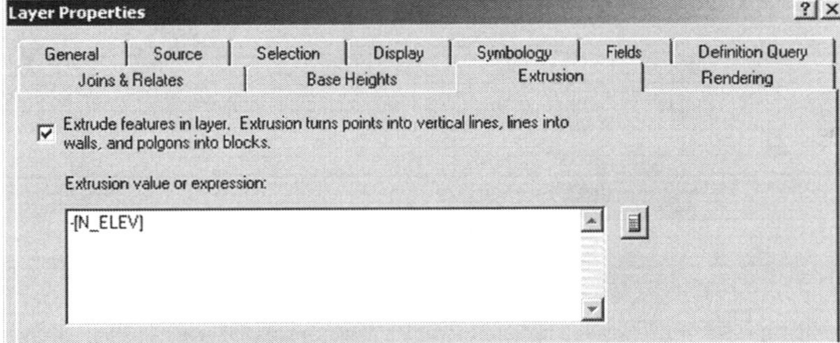

Click the Rendering tab in the Layer Properties dialog. Check the box to use smooth shading. When polygon or line features are extruded, this box is unchecked by default. (When point features are extruded, no shading is available.)

Click OK to close the Layer Properties dialog.

Now the base heights are set for both layers, and the mines are extruded. Before you take a look down under, let's give the raster some hillshading to for better contrast against the mines.

Step 11. Make a Hillshade of Slo_cnty

This step will give you a preview of some of the surface analysis you'll be doing in later chapters.

If the 3D Analyst menu isn't already visible, right-click anywhere in the ArcScene menu area to bring up the Toolbars menu, then check the box next to 3D Analyst.

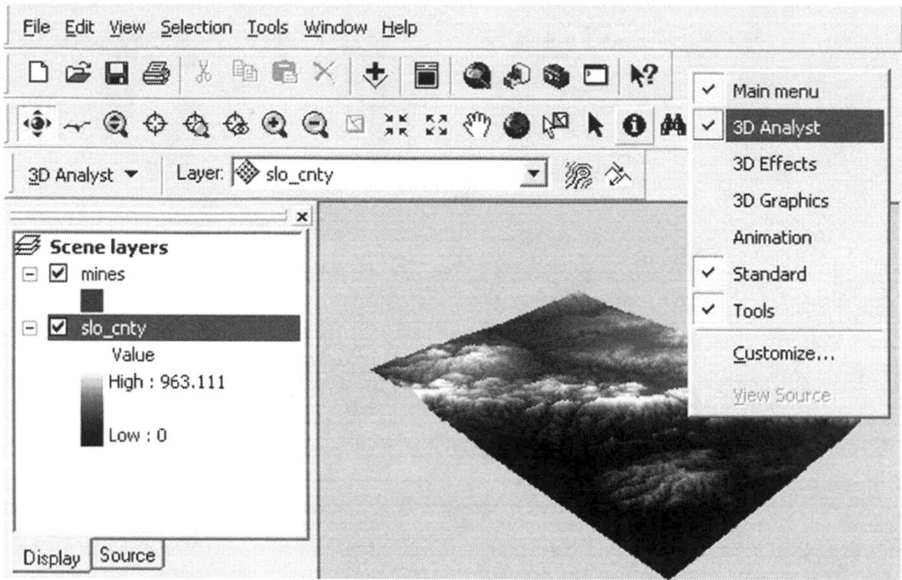

Click the 3D Analyst dropdown menu in ArcScene. Select Surface Analysis, and then click Hillshade.

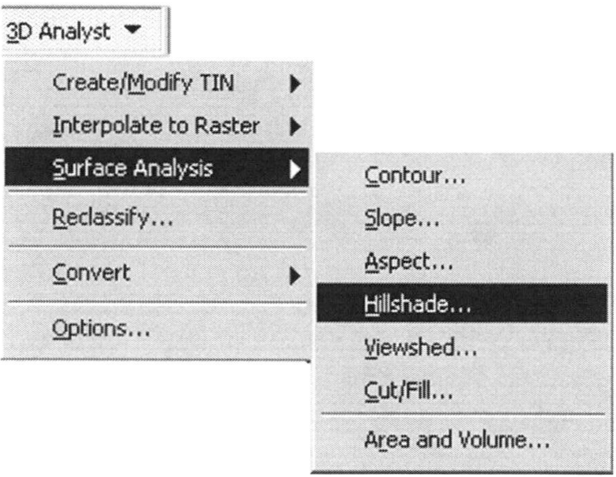

In the Hillshade dialog, the input surface is slo_cnty.

Click the Browse button next to the Output raster box, and navigate to your 3DDATA\Chapter02\MyData folder.

Name the new layer hshd_slo.

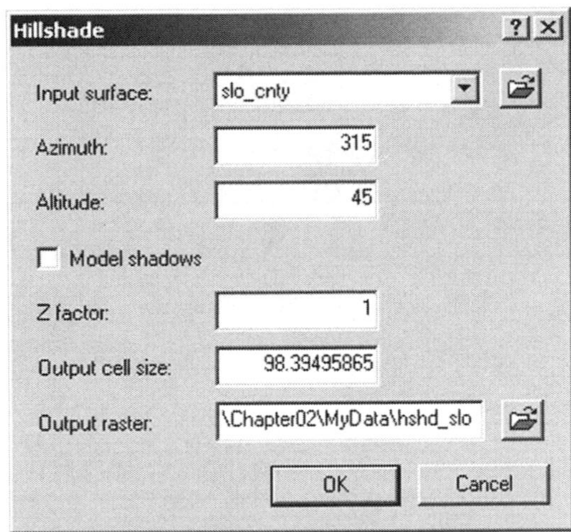

Click OK, and wait for the hillshade to be processed. When it's finished it will be added to the scene.

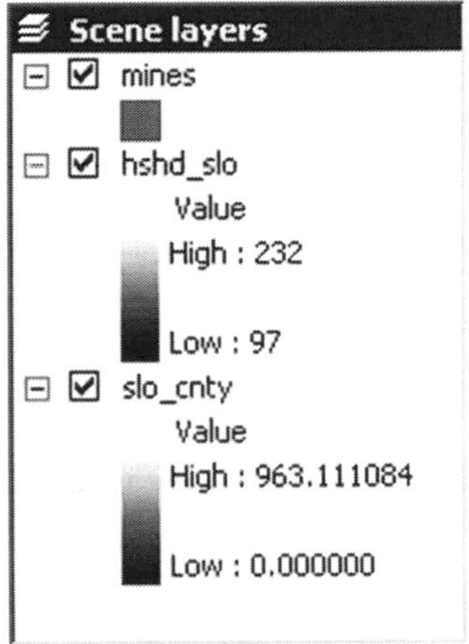

Step 12. Set Base Heights for the Hillshade of Slo_cnty

In the Table of Contents, right-click hshd_slo and choose Properties. Click the Base Heights tab, and check "Obtain heights for layer from surface."

From the dropdown list, choose slo_cnty (not hshd_slo) as the source for base heights.

Click OK to close the Layer Properties dialog.

In the Table of Contents, turn off slo_cnty so that it won't compete for display space with hshd_slo.

Use the Navigate tool to zoom into the scene, and take a look underneath the hillshade layer. The mines are extruded downward by the values in the N_ELEV field in their attribute table. (The units are feet.)

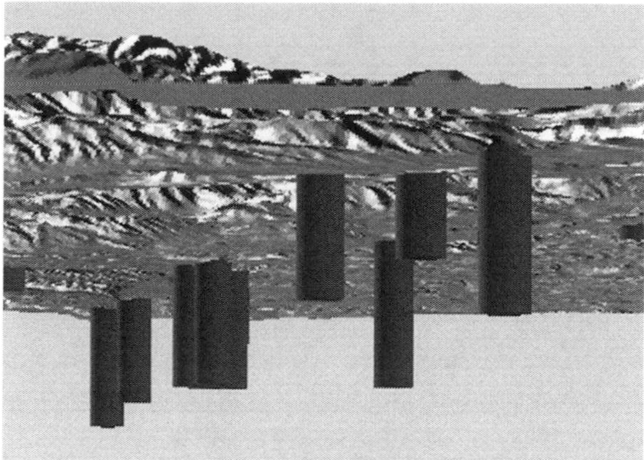

The vertical exaggeration of the scene affects the mine depths as well as the raster heights. Without such a large vertical exaggeration, the mines would barely poke beneath the surface.

Step 13. Symbolize the Mines by Type

Let's give the mines a bit more personality by coloring them according to their type.

In the Table of Contents, right-click mines, choose Properties, and click the Symbology tab.

In the Show box, choose Categories: Unique values.

In the Value Field dropdown list, choose MINETYPE.

Click the Add All Values button. Uncheck the box next to the topmost symbol that says "All other values."

Choose a style that you like in the Color Scheme box.

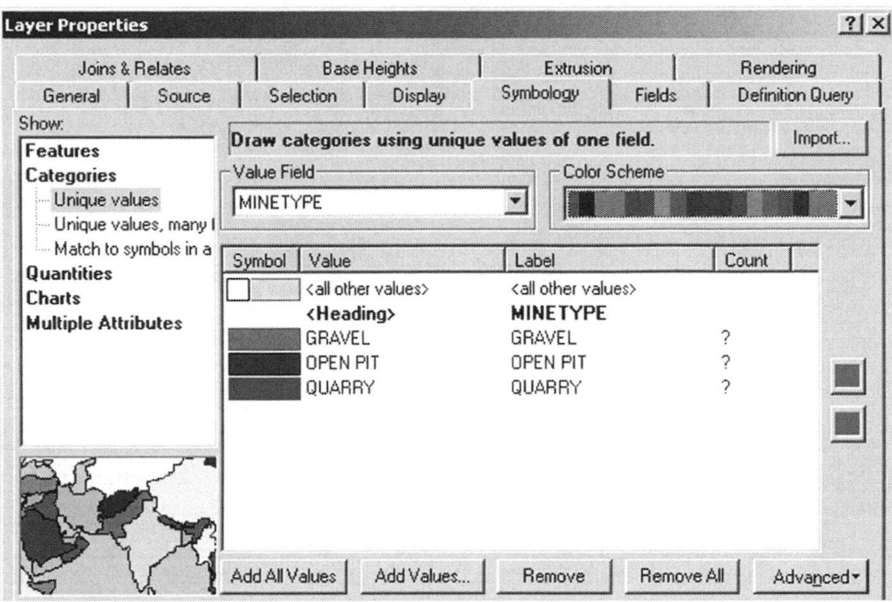

Click OK to close the Layer Properties dialog.

Now the mine polygons are extruded by depth and symbolized by type. Take a look around and examine your work.

Step 14. Close ArcScene

When you're finished checking out the mines, close ArcScene. If you'd like to save the document, give it a name of your choice and save it in your 3DDATA\Chapter02\MyData folder.

Challenge Exercise

View Regional Park Study Data in ArcScene

This is the first of several challenge exercises involving an imaginary regional park that we'll be developing using ArcScene, ArcGlobe, SketchUp, and Google Earth. In this exercise, you'll take a first look at the datasets for the project.

Step 1. Start ArcScene and Add the Data Layers

You'll be working with six data layers within the Haddix topographic quadrangle in eastern Kentucky, an area located between the towns of Jackson and Hazard.

Double-click the ArcScene icon on your desktop, or open it from the Start menu.

Click the Add Data button, and navigate to your 3DDATA\Chapter02\Data folder. By holding down the Cntrl key while selecting with the mouse, you can add several layers at once. So add the following datasets to the scene:

o51_dem

o51_hshd

o51_landslide.sid

elk_park_bnd.shp

haddix_rds.shp

haddix_creeks.shp

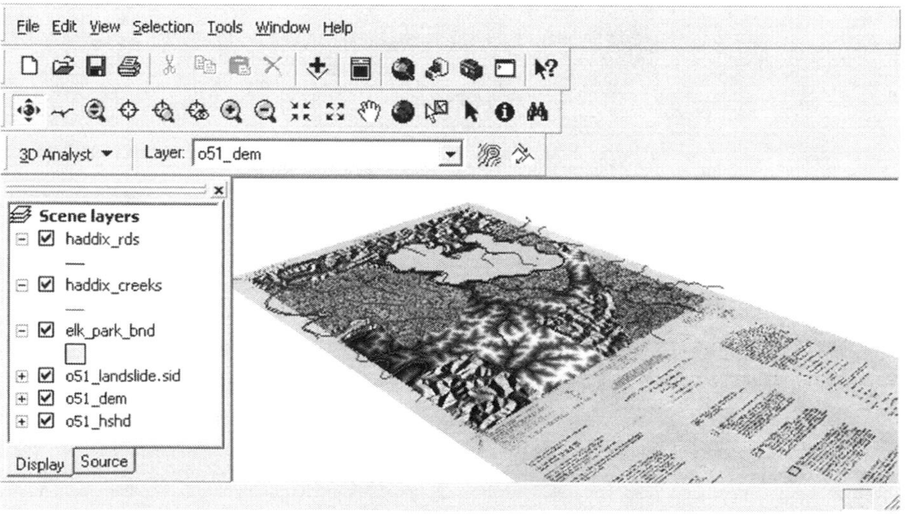

We have a nice collection of roads, creeks, a DEM, a hillshade model, and a MrSID image of the landslide activity in the Haddix topoquadrangle. Granted, the landslide map was last updated in 1979, but it will do for the exercises in our imaginary park model. We also have a polygon created by following Troublesome Creek and Lost Creek around the approximately nine-square-mile area that makes up our park.

Examine these layers, taking note of the attribute tables for haddix_creeks, haddix_rds, and the elk_park_bnd shapefiles. Also check out the MrSID layer. If the resolution is too coarse, you can improve it by going to Layer Properties, clicking on the Rendering tab, and setting the "Quality enhancement for raster images" to High. Depending on how powerful your processor is, however, this can significantly slow down drawing speed, so you might want to keep most of your raster layers at a medium quality when navigating, especially with multiple datasets in the scene.

Step 2. Set Base Heights for the Layers

Now that you've had practice setting base heights for several types of data, try setting the base heights for all six layers in ArcScene. Remember, to set base heights for a layer, right-click it in the Table of Contents, choose Properties, click the Base Heights tab, and click "Obtain heights for layer from surface." The layer you'll use for base heights in each case will be o51_dem.

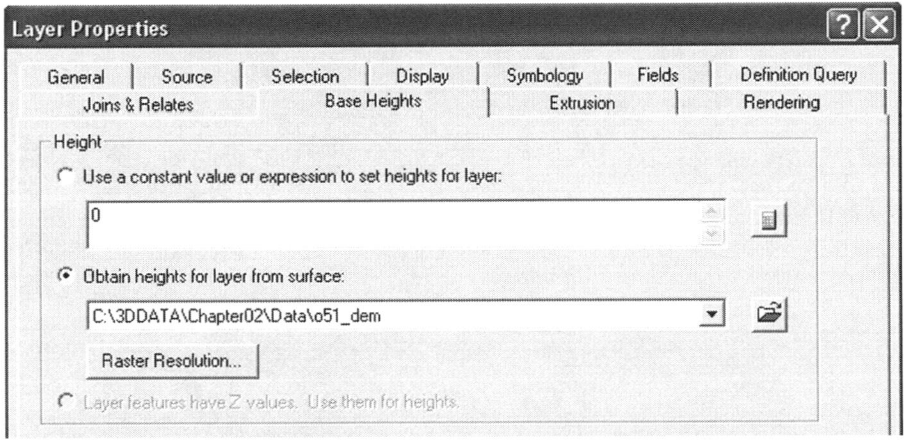

Use o51_dem as the source for base heights for each of the six layers in ArcScene

Notice that when you set the base heights for the MrSID image, the cells in the image are not only raised to the elevations in the DEM but also cropped to the extent of the DEM. This is generally desirable, but there are times when it isn't the result you want. There are various ways around it, such as adding a layer twice to the Table of Contents and only setting the base heights for one instance of the layer.

Notice further that the elk_park_bnd doesn't do very well when set to the base heights of the DEM. The polygon's vertices exist only along its outer perimeter, so only the outer perimeter is given z values from the DEM. The interior elevations are interpolated from the elevations along the edge. This is similar to the "stitching" problem we saw earlier in Exercise Six.

In a later lesson you'll learn how to convert the polygon to a multipatch feature to solve this problem, but another way around it is to symbolize elk_park_bnd with a heavy outline and no fill color.

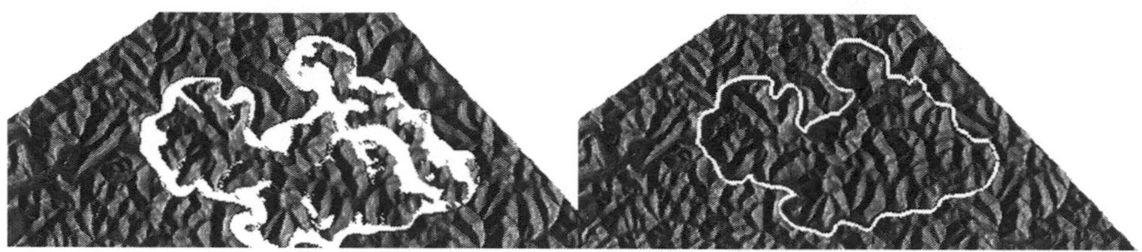

Symbolizing the Elk Park polygon with a heavy outline instead of a fill makes for a better 3D display

Step 3. Change the Drawing Priority, Symbology, and Transparency of the DEM and the Hillshade

Change the color ramp used to symbolize the DEM's elevation in the Symbology tab. If you still have the 3D Effects toolbar loaded, you can set a transparency for the DEM that lets you see the hillshade through it. It's best in that case to set the hillshade's drawing priority to something other than "1" so that it won't compete for space with the DEM. Remember that you can use the Rendering tab in the Layer Properties dialog of the hillshade to change its drawing priority.

On the 3D Effects toolbar, increasing the Transparency for o51_dem allows the hillshade to be viewed through the DEM, to give the impression of more terrain definition

Step 4. Set Vertical Exaggeration and Background Color for the Scene

In the Table of Contents, double-click Scene Layers to bring up the Scene Properties. Click the General tab. In the Vertical Exaggeration box, choose 2.

In the Background color box, choose Apatite Blue. Click OK. Click the Navigate tool, zoom in, and take a look around the scene.

Step 5. Close ArcScene

When you're finished familiarizing yourself with all the layers, close ArcScene. If you'd like to save the document, give it a name of your choice and save it in your 3DDATA\Chapter02\MyData folder.

In this chapter you learned how to set 3D display properties for scenes and layers. In Chapter 3, you'll learn how to animate your data, fly through terrain, and make movies.

CHAPTER **3**

3D Navigation and Animation

Targets and Observers

You already know how to use the Navigate tool to pan, zoom, and move around your data in a 3D scene. In this chapter you'll learn to specifically control your perspective in a scene, fly through terrain, create an animation sequence, and save it as a movie file.

When you change your perspective with the Navigate tool, you are telling ArcScene to change the position of the target (what you're looking at) and the observer (you). Officially, the observer is the point of the scene viewer relative to the data in the scene, and the target is the center point around which a scene viewer moves when you navigate. The relative position between the target and observer defines your 3D perspective.

Target and observer points each have an x, y, and z value. You can set specific target and observer locations, either by clicking in the scene with the Target and Observer tools, or by typing x, y, and z coordinate values in the scene's View Settings.

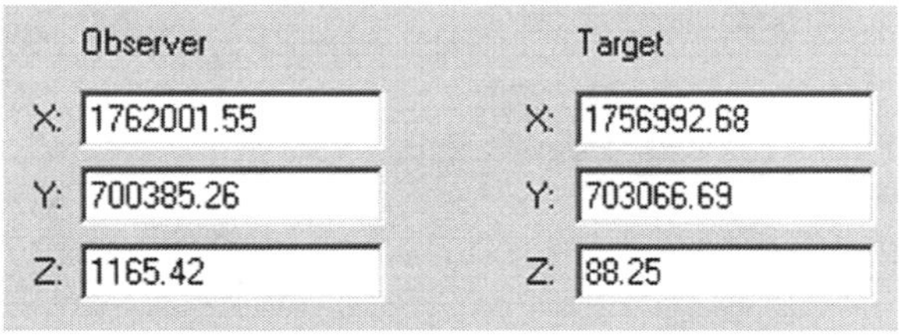

	Observer		Target
X:	1762001.55	X:	1756992.68
Y:	700385.26	Y:	703066.69
Z:	1165.42	Z:	88.25

The star indicates the location of the target in the scene. The x, y, and z values of both target and observer are shown in the box above

Animated Rotation and Flight Simulation

You can also make your data spin around its axis at various speeds, while you pan, zoom in and out, or navigate. Rotating a scene is a good "hands-off" way to get an overview of the data.

The Fly tool lets you navigate through data from a bird's-eye view. When the Fly tool is activated, the target and observer coordinates change continually, giving the illusion of flight over a motionless landscape.

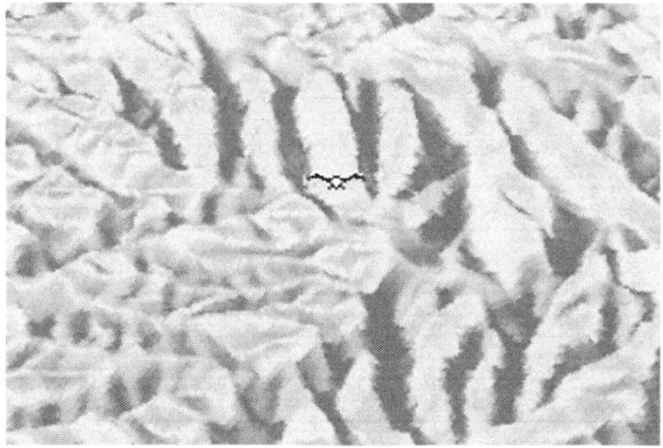

Flight simulation through the terrain of a hillshade map

Creating Animation

After you've had some experience flying through your data and placing targets, you'll create an animation file by linking scenes together. ArcScene lets you make movies that show changes in scene properties (illumination, background) perspective (navigation, rotation, flight) or any change that you apply to your data. These films can be saved as ArcScene animation documents or converted to .avi files.

An .avi file of ArcScene animation

Exercise 3-1

Set Targets and Observers

When you're looking at data in ArcScene, you might want to know what you can see from a specific position, or what an object looks like from different angles. In this exercise, you'll learn about targets and observers, orthographic and perspective view, roll angle, and pitch.

Step 1. Start ArcScene and Add Data

Open ArcScene from your desktop, and click the Add Data button.

Navigate to your 3DDATA\Chapter03\Data folder. Add 3d_bldgs.lyr and tar_tin.lyr to the scene. (If you're using version 9.1 or 9.2, add 3d_bldgs_v91.lyr and tar_tin_v91.lyr.)

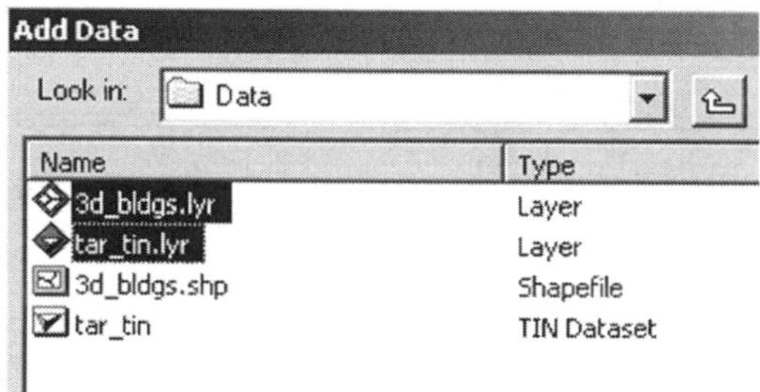

You've added a TIN layer and a layer of buildings to the scene. One of the features is a bright yellow 20-foot pole representing an observation tower, from which you'll be surveying the scene.

Step 2. Add a Background Color

In the Table of Contents, right-click Scene Layers and choose Scene Properties.

In Scene Properties, click the General tab. Click the dropdown box next to Background color, and choose Sodalite Blue.

Click OK.

Step 3. Examine the Scene

Use the Zoom In and Navigate tools to take a look at your surroundings.

Step 4. Center on a Target

Click the Full Extent button, and then zoom in to the cluster of blue and red buildings.

Click the Center on Target tool. 

This tool centers the view on any point that you click in the scene. Click on the roof of the red building as indicated in the graphic below.

Notice that the scene pans to make the rooftop the center (or near center) of the scene.

Step 5. Navigate Around the Set Target

Click the Navigate tool.

Hold down the left mouse button, and move the cursor around in the scene. Notice that the surface rotates around the point you set as the target.

Now hold down the right mouse button, and drag the cursor up and down in the scene. Notice that the target remains the focus while zooming. (The only way to change the target is to hold down both mouse buttons, which will pan to a different location).

You can also center on a target while you're using the Navigate tool. To do so, hold down the Ctrl key, and then left-click anywhere in the scene. The scene is centered on the new location.

Step 6. Zoom to a Target

Click the Full Extent button. The target is reset to the center of the surface.

Click the Zoom to Target tool. 

This tool lets you set and zoom in to a target with one click. Click the rooftop of the red building again, as best you can with the scene set to its full extent. ArcScene zooms in to the top of the building.

If you keep using the Zoom to Target tool while you're this close, you'll lose sight of your target. The Zoom to Target tool works best from a greater distance.

Click the Navigate tool, and click the Full Extent button again. This time, zoom to a target while you're using the Navigate tool. To do so, hold down the Ctrl key, and then right-click on the desired spot. The scene will center and zoom in on the new location.

Step 7. Set an Observer

Click the Full Extent button. Use the regular Zoom In tool to make your scene looks like the following graphic:

You're going to target the rooftop again, but you'll also set a point of observation.

Click the Center on Target tool. 

Click the red building. The scene is centered on the rooftop. Make sure that you can still see the yellow pole on the facing slope of the hill. (If you can't, use the Full Extent, Zoom In, and Center on Target tools again until you can see the red building and the yellow pole.)

Now click the Set Observer tool. 

Click on the yellow pole on the hillside. ArcScene zooms in and relocates your point of observation. You should be looking at the buildings, now, from approximately the following perspective:

If you don't get a point of view close to the one above, run through Step 7 again. It takes a little practice and a little luck.

Step 8. Set Target and Observer Coordinates

Clicking target and observer points in the scene is quick and convenient, but if you know their coordinates you can create a more exact model of the visual landscape. The View Settings dialog box allows you to control x, y, and z positions of both observer and target precisely, and reports the line-of-sight distance between them.

Click the Navigate tool, and click the Full Extent button to return to the original scene.

Open View Settings from the View menu.

In it you see the coordinates of the current observer (you) and target (the middle of the scene). You also see that it's in Perspective or 3D view, and that the roll angle and pitch are set to 0 and 29.

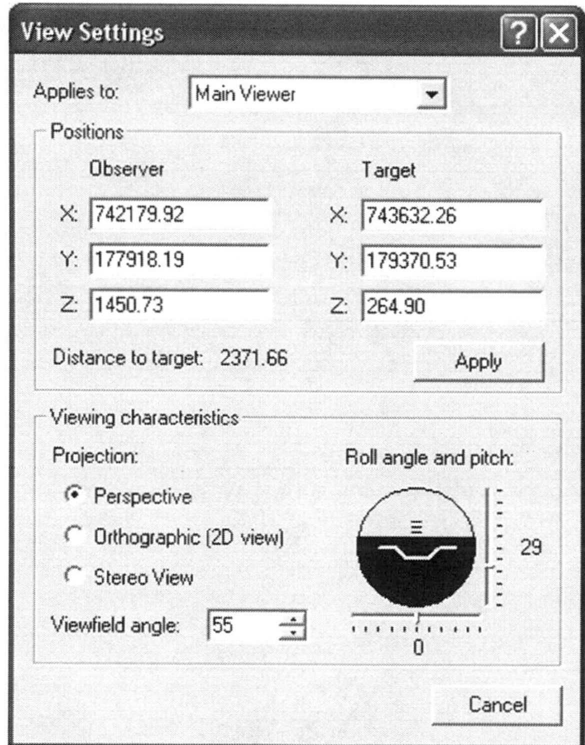

Step 9. Change Target and Observer Coordinates

Move the View Settings dialog so that you can see the contents of the scene.

Type the following coordinates into the Observer and Target fields, so that the dialog looks like this:

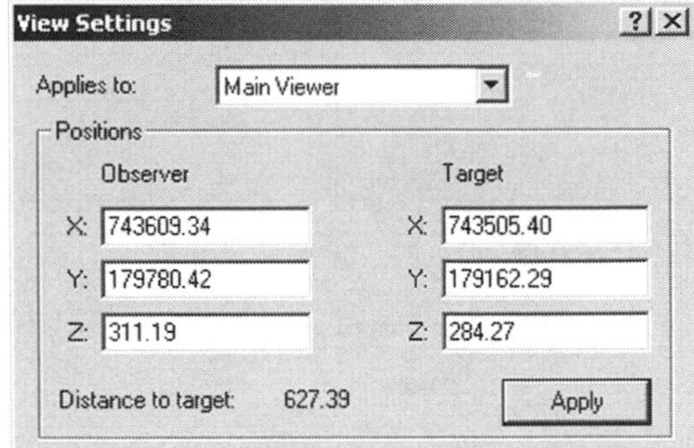

When you're finished typing in both sets of numbers, click Apply.

3D Analyst calculates the values, zooms to the observer position and faces the target location.

The target and observer coordinates that you typed in are the same locations that you've been working with. The resulting scene should look pretty familiar:

Notice that the distance from your perspective—the top of the yellow pole—to the rooftop of the closest red building is about 627 meters. This is a 3D measure, which means that it accounts for vertical as well as horizontal distance. There is a direct line of sight between target and observer. If the pole were a cell tower and you were to climb to the roof of the building to talk on your cell phone, you would probably get excellent reception.

Step 10. Create a Bookmark

In the next few steps, you'll be moving the scene around, and you may lose the current perspective. Since you probably don't want to type in the target and observer values all over again, now's a good time to set a bookmark.

From the Bookmarks menu, choose Create. (If you're using an earlier version of ArcScene, click the View menu, then the Bookmarks menu.)

Since you're only making one bookmark, you can leave the title "Bookmark 1." To get back to this view at any time, click the Bookmarks menu.

Click OK to close the 3D Bookmark dialog.

Step 11. Adjust Angle and Projection

The View Settings dialog should still be up. If not, open it again from the View menu. Drag it out of the way of the scene.

Look at the fields in the Viewing Characteristics panel. The projection is Perspective (3D), and the Viewfield angle is around 55 (yours may be different).

In ArcScene, experiment with the Narrow Field of View and Expand Field of View buttons. 

The number in the Viewfield angle box changes each time to reflect the new distance. Narrow the field of view to 25.

You can also type values directly into the Viewfield angle box, or use the up and down arrows. Type in the value "60." The scene pulls back to the larger viewing angle.

Step 12. Change Projection

Under Viewing Characteristics, click Orthographic.

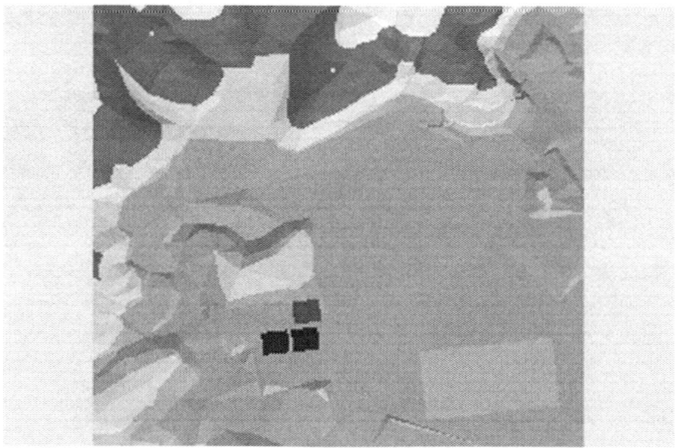

ArcScene shows the full extent of the scene in two dimensions, just like ArcMap. Orthographic scenes are limited to 2D navigation, so many of your 3D tools will no longer work. The Narrow/Expand Field of View buttons, the Set Observer tool, the Navigation tool, and the Fly Tool are disabled.

You can still use the Center on and Zoom to Target tools, the Pan tool, and the Zoom In/Out tools to get a closer look. When you're done, click the Perspective button.

Step 13. Change Roll Angle and Pitch

You can also adjust the pitch to elevate your point of observation. If observer and target are at the same height, the pitch is zero. If the observer is higher, the pitch is positive; if the observer is lower, the pitch is negative.

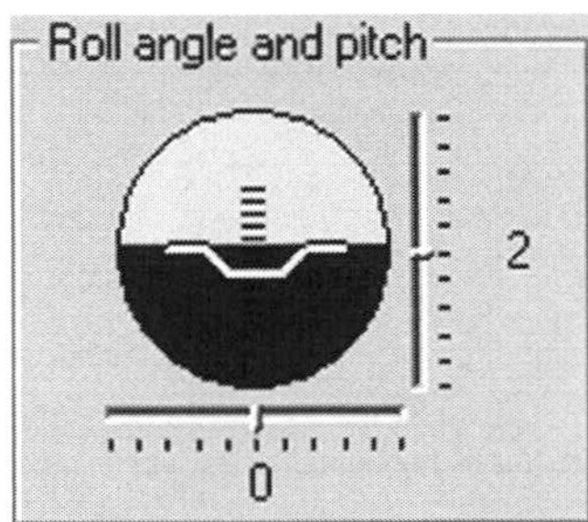

Select "Bookmark 1" from the Bookmarks menu.

Make sure that you're using Perspective view in the View Settings dialog. Drag the vertical Pitch slider up and down to change your elevation with respect to the target. The pitch ranges from +89 to −89 degrees. At +89, your observation point is almost directly above the building. At −89, it's underneath the surface of the TIN.

Slide the vertical bar back to 1 or 2 degrees.

Now slide the horizontal bar to the left and right. This controls the roll angle, and is similar to looking out of the cockpit of a plane while it tips its wings left and right. This angle ranges from 180 to −180 degrees.

Step 14. Close ArcScene

Feel free to experiment with the View Settings dialog from different perspectives. When you're finished, close ArcScene. If you'd like to save the scene, give it a name of your choice and save it in your 3DDATA\Chapter03\MyData folder.

Exercise 3-2

Animated Rotation and the Viewer Manager

Animated rotation puts your data in motion so you can look at it without using the mouse or keyboard. In this exercise, you'll animate a TIN surface from San Luis Obispo, CA. You'll also use the Viewer Manager to look at the surface from three simultaneous perspectives.

Step 1. Start ArcScene and Add Data

Open ArcScene, and click the Add Data button. Navigate to your 3DDATA\Chapter03\Data folder. Add hill_tin to the scene.

Step 2. Change the TIN's Symbology

In the Table of Contents, double-click hill_tin to bring up the Layer Properties dialog.

Click the Symbology tab.

In the Show box, Edge Types is checked. Highlight "Edge Types," and click the Remove button under the Show box. Also remove "Faces."

Click Add to open the Add Renderer, and choose "Face elevation with graduated color ramp." Click Add again, then Dismiss to close the Add Renderer.

Back in the Layer Properties dialog, right-click the dropdown arrow next to the Color Ramp list and uncheck "Graphic View" so that you can see the names of the color ramps.

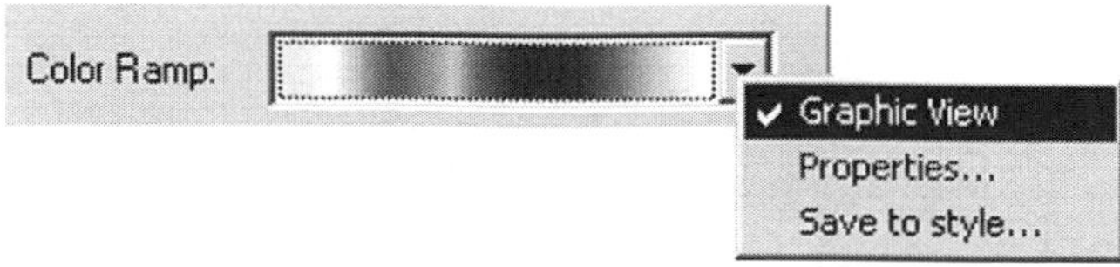

Choose the ramp called "Elevation # 2."

Click OK to dismiss the Layer Properties dialog.

Step 3. Enable Animated Rotation

Click the Navigate tool, and take a look at the surface. Use the Identify tool to examine the attributes of the hill.

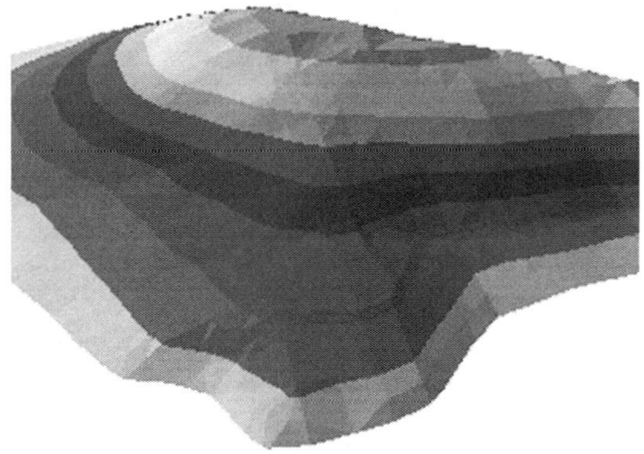

When you're finished navigating around the TIN, click the Full Extent button.

In the Table of Contents, right-click Scene Layers and choose Scene Properties.

Click the General Tab.

Check the box next to Enable Animated Rotation.

☑ Enable Animated Rotation

When you use the Navigation tool to rotate the scene, hold down the left mouse button, drag in the direction you want the scene to rotate, and release the mouse button while the scene is moving.

Click OK.

Step 4. Start Animated Rotation

Click the Navigate button if it isn't already selected, and place your cursor at the right side of the display. You'll notice that when animated rotation is active, the Navigate tool's cursor is black and white with a circle around it.

Hold down the left mouse button, drag the cursor to the left, and release the mouse button while you drag it. There's a bit of an art to this: if you don't release the mouse button until you are well to the left side of the display, ArcScene will treat your command like regular navigation. Try to release the mouse button at the midpoint of the display.

Once the TIN is spinning, take your hand off the mouse.

Step 5. Set the Target and Observer During Animated Rotation

If the TIN is rotating too quickly or too slowly, use the Page Up or Page Down key to increase or decrease its speed. (Your cursor has to be in the scene for the Page Up and Page Down buttons to affect the rotation speed.) The Arrow keys can be used to affect the tilt of the surface.

The scene rotates when the Navigate tool is active, but you can still use the other tools on the 3D Analyst toolbar. When you click one of them, the rotation is temporally suspended. To restart the rotation, just click the Navigate tool.

While the TIN is revolving, hold down the right mouse button and zoom in.

Now click the Center on Target tool. 

The rotation stops. Click on one of the lower elevation bands on the side of the TIN, as indicated in the graphic below.

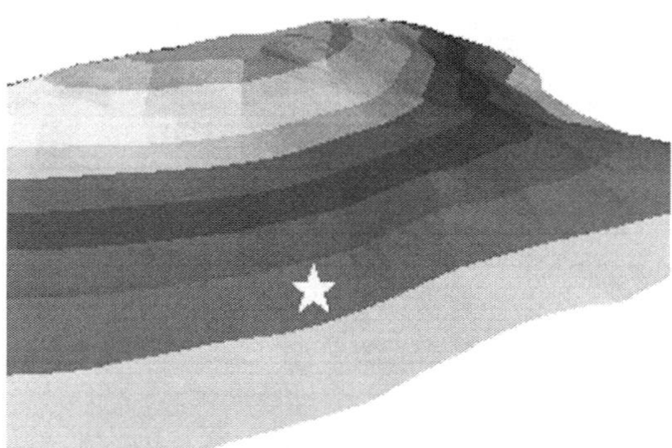

The target becomes the center of the scene.

Click the Navigate tool again, and the rotation continues, centered around the new target.

Now click the Set Observer tool. 

Click on one of the upper elevation bands of the TIN. Try to choose a location that has a direct line of sight to the target.

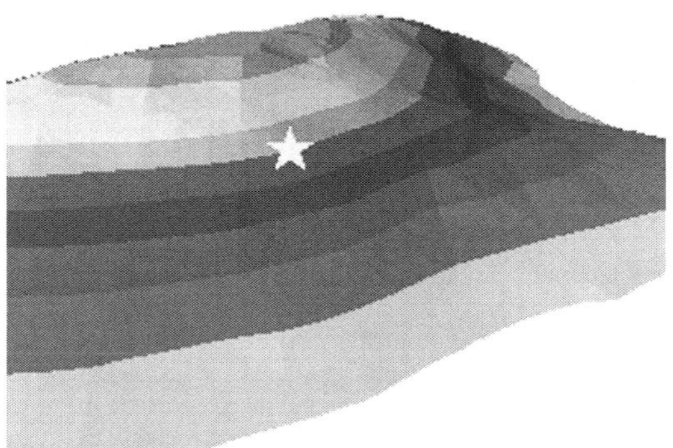

The scene zooms to the new location and then stays motionless while you look down at the target from the observation point.

Click the Navigate tool again. The scene now rotates 360 degrees around the target, starting from the new observer's position.

Step 6. Stop Rotation

To stop rotation at any time, press the Esc key while the cursor is in the display.

To resume rotation, hold down the left mouse button and drag the cursor to the left or right, releasing the mouse button as you pass the midpoint of the display.

While the TIN revolves, click the Full Extent button. Experiment with some of the other navigational tools.

When you're finished, press the Esc key to stop the rotation. Notice that the Esc key only works when the cursor is in the display.

Step 7. Turn Off Animated Rotation

Double-click Scene Layers in the Table of Contents, and under the General tab uncheck Enable Animated Rotation.

Click OK to dismiss the Scene Properties dialog.

Step 8. Add a Viewer

By adding viewers, you can compare several perspectives of one scene side by side.

Click the Full Extent button.

Click the Add New Viewer button. 

A new display appears, titled "Viewer 1."

Move Viewer 1 out of the way of the ArcScene application window.

Click the Navigate tool if it is not already selected, and use it to match Viewer 1 to the following graphic.

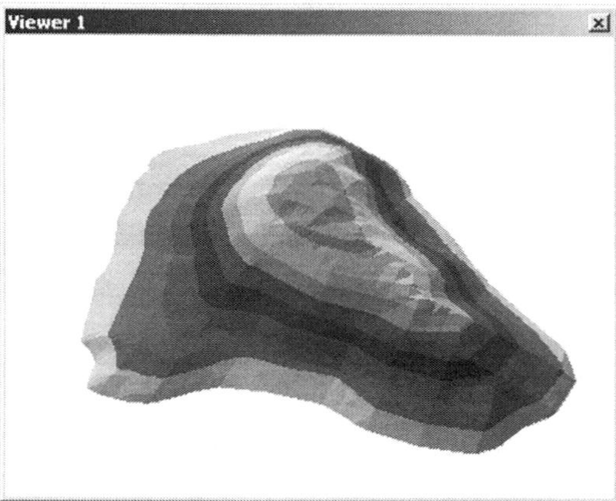

The navigational tools work in the new viewer as well as in the original display. As long as Viewer 1 is active, the tools in ArcScene will apply to it.

Step 9. Add Another Viewer

Click the Add New Viewer button again. 

Another display appears, titled "Viewer 2." Notice that whenever you add a new viewer, it starts out at the full extent of the data.

Move Viewer 2 so that you can see all three scenes. Use the Navigate tool to steer and zoom in to Viewer 2 until it looks something like this:

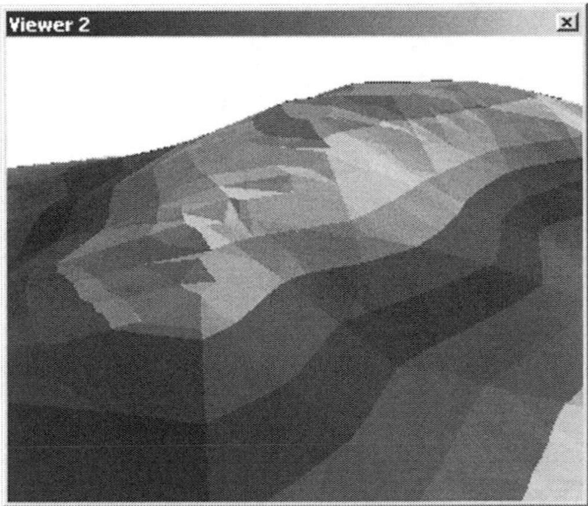

Step 10. Apply Scene Properties to All Three Viewers

It might be helpful at this point to distinguish between layer properties, scene properties, and view settings. You know that you can change individual layer properties, such as symbology and base heights, by clicking on a layer in the Table of Contents and bringing up its Layer Properties dialog. Scene properties—illumination, vertical exaggeration, coordinate system, background color—affect all layers equally in a scene, and, in fact, all viewers of that scene. View settings, on the other hand, such as target and observer positions, orthographic and perspective view, and roll angle and pitch, pertain to navigation and can be tailored for each view separately.

In this step, you'll change background color and illumination (both are global scene properties). In the next step, you'll change the view settings for two of the scenes.

From the View menu in ArcScene, choose Scene Properties. Move the Scene Properties dialog so that you can see all three viewers.

Under the General tab, click the Background color dropdown arrow and choose Yogo Blue.

Click Apply. The sky turns blue in all three viewers.

Now select the Illumination tab.

With your cursor, adjust the azimuth and the altitude of the sun. Again, all three viewers reflect the changes.

Type "225" into the azimuth box, and "30" into the altitude box.

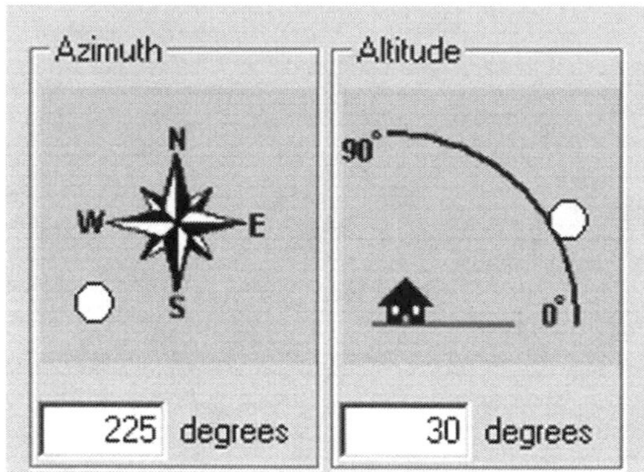

Click OK to dismiss the Scene Properties dialog.

Step 11. Change View Settings

From the View menu choose View Settings. Move the View Settings dialog so that you can see all three viewers.

Click the dropdown arrow next to "Applies to" at the top of the View Settings dialog.

Settings are applied to one viewer at a time. Choose Viewer 1.

In the Viewing Characteristics panel, choose Orthographic. This gives you a top-down view of the hill in 2D. Notice that in the Positions panel, the Observer's z-position and the Target's x-, y-, and z-positions are grayed out.

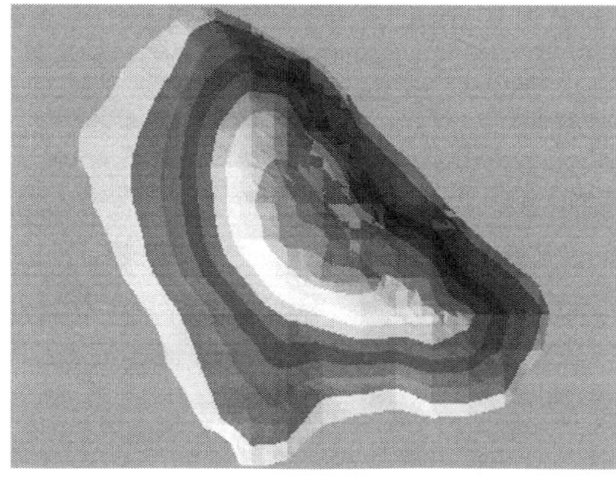

Now choose Main Viewer from the "Applies to" dropdown list.

In the View Settings dialog, change the Roll Angle from 0 to −35.

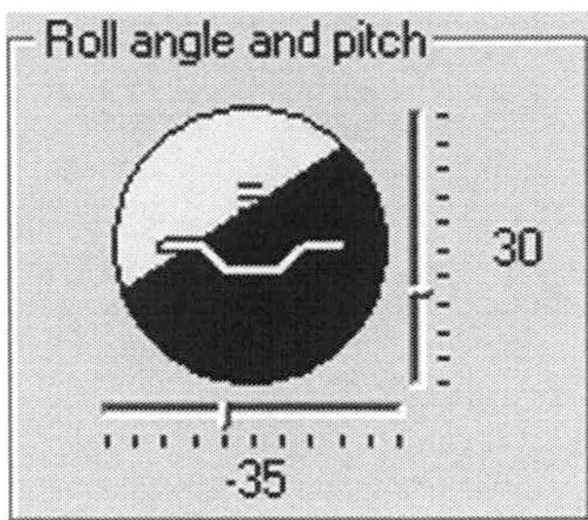

Close the View Settings dialog box.

Now double-click Scene Layers in the Table of Contents to bring up the Scene Properties dialog.

Click the General tab, and check Enable Animated Rotation.

Click OK.

Make sure that the Navigate tool is selected. The cursor changes to let you know that animated rotation is active.

Place the cursor over the Main Viewer, hold down the left mouse button, and swipe it either direction across the scene, releasing it somewhere toward the middle.

Use the Page Up and Page Down keys to speed up or slow the rotation.

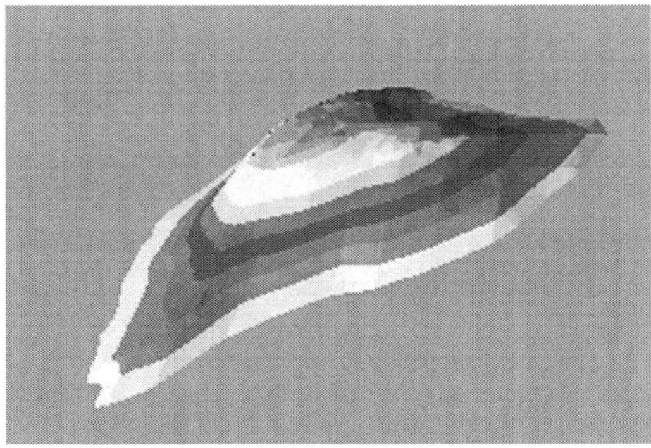

Now hold down the left mouse button and swipe the cursor over Viewer 2. If you don't have too many applications open and you have a powerful computer, you can set it spinning as well.

Press the Esc key to stop the rotation in Viewer 2. Put your cursor in the Main Viewer, and press the Esc key to stop its rotation.

You may notice a couple of things: One, if you put your cursor over Viewer 1, it looks like animated rotation is available, but Viewer 1 is in orthographic perspective, so rotation in that viewer won't work. Two, even though Animated Rotation in enabled from the Scene Properties dialog, it is not a global property for all scenes. Since you activate it with the mouse, you can apply it individually to any viewer.

Step 12. Manage Viewers

From the Window menu, choose Viewer Manager.

The Viewer Manager lets you rename, show, hide, restore, or close any viewers other than the main ArcScene application. If you were to click Close Viewer(s) at any time, you would delete the viewer that is highlighted. Likewise, if you were to click the X in the upper-right corner of a viewer, you would delete it. The Restore option applies to a viewer that has been minimized or maximized, not closed. In this case, we want to keep our viewers, so don't click those options.

With "Viewer 1" highlighted, left-click once on the name. A text box with a blinking cursor appears around it, inviting you to type over the name. Do so, and rename it "Ortho_view."

Press the Enter key to confirm the new name.

Highlight Viewer 2, and rename it "Hilltop_view."

The two viewers now have more descriptive titles.

With Ortho_view selected, click Hide, then Show, then Hide. Keep in mind that this is different from selecting "Close Viewer(s)."

Click OK to dismiss the Viewer Manager.

Hilltop_view should still be visible. Double-click the middle of its title bar to maximize it.

Now double-click the title bar again to return it to normal size.

Right-click the title bar, and take note of the options for creating a bookmark, and for maximizing, minimizing, restoring, and deleting the viewer.

Step 13. Close ArcScene

Experiment as much as you like with View Settings, Scene Properties, and the Viewer Manager. When you're finished, close ArcScene. If you want to save the scene, give it a name of your choice and save it in your 3DDATA\Chapter03\MyData folder.

Exercise 3-3

The Fly Tool

Now that you've had a chance to use animated rotation, you'll test your wings with the Fly tool, which lets you control your speed, altitude, and direction as you navigate through a landscape.

Step 1. Start ArcScene and Add Data

Open ArcScene, and click the Add Data button. Navigate to your 3DDATA\Chapter03\Data folder.

Add flight_path.shp and tar_tin.lyr to the scene. (For version 9.1 or 9.2, add tar_tin_v91.lyr.)

You used tar_tin.lyr in Exercise 1 when you learned how to set targets and observers. Since it's fairly small with both flats and hills, it makes good terrain for flight practice.

Step 2. Symbolize the Flight Path

In the Table of Contents, click the line symbol under flight_path to bring up the Symbol Selector.

Choose the Highway symbol. Click the Color dropdown arrow in the Options box and choose a bright yellow.

Click OK to close the Symbol Selector dialog.

Step 3. Look at the Shapefile's Attribute Table

The flight path was drawn as a 2D polyline shapefile in ArcMap, and then converted to 3D using tar_tin as the elevation source.

Right-click flight_path and choose Open Attribute Table.

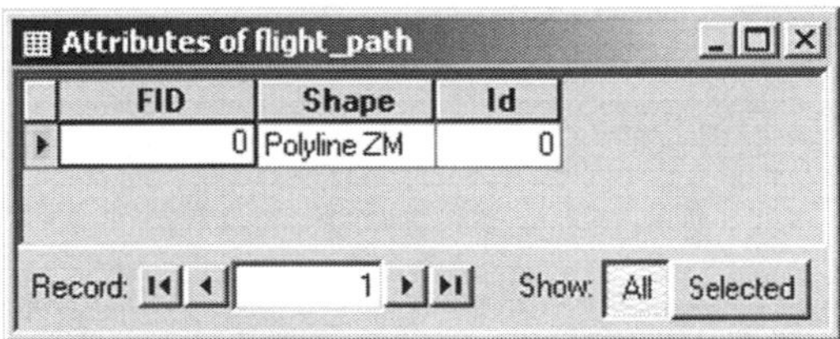

The Z in the Shape field indicates that this is a 3D shapefile. You could also use a 2D shapefile and set its base heights to tar_tin, but converting it to 3D saves a step if you share your 3D data or use the same datasets regularly. You'll learn more about 3D shapefiles in later chapters.

Close the attribute table.

Step 4. Use the Fly Tool

The Fly tool takes a certain amount of practice—more for some, less for others—but in any case, you'll need to memorize the commands to use it.

Fly Tool Instructions	
Action	**Command**
Activate Fly Tool	Click the Fly Tool, then left-click on the scene
Start flight	Left-click
Increase speed	Additional left-clicks
Reduce speed	Right-click
Stop flight	Press the Escape key, or click the middle mouse button

Some other important facts:

You can fly backward by starting your flight with a right-click. Further right-clicks will increase your backward speed; left-clicks will reduce it.

The Shift key is invaluable when using the Fly tool. Holding down the shift key while flying forward or backward maintains your altitude, and keeps you from crashing into the surface or flying off into space.

To fine-tune your travel speed, press the arrow up or down keys in between left- or right-clicks of the mouse.

Click the Fly tool. 

Put your cursor over the scene. It appears as a bird standing on a cloud until you click it.

Click the left mouse button once, and look down at the status bar. (If you don't see the status bar in the lower-left corner of ArcScene, click the View menu and check Status Bar.)

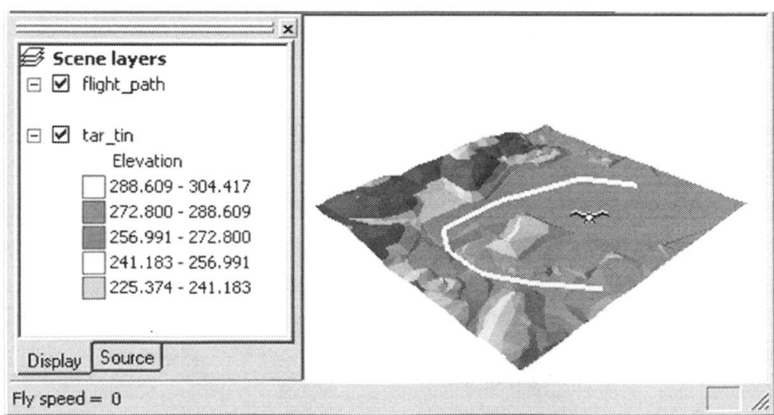

At this point, your fly speed is set to zero, and you can change the direction you're facing before starting the fly through. Once you're in motion, flight through the scene will follow the movements of the mouse.

Using the commands listed on the Fly Tool Instructions table, practice flying over the surface in different directions and speeds. You may want to use the Navigate tool and the Zoom In tool to get your bird to a good starting position.

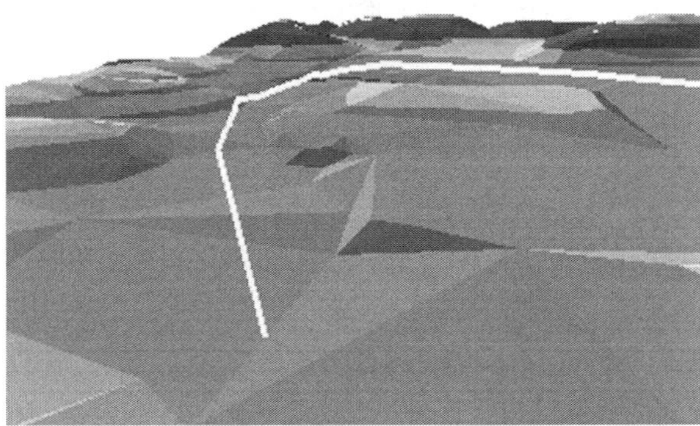

Work on it until you feel somewhat comfortable controlling your flight. Try flying forward along the yellow line, then backward along it. If you lose control, press the Esc key, reorient yourself with the Full Extent button, and try again. Don't worry if you aren't an expert right away; like riding a bicycle, it takes a few sessions.

Step 5. Add Another Flight Path

Click the Add Data button, and add flight_path2.shp

Step 6. Symbolize the New Flight Path

In the Table of Contents, click the line symbol under flight_path2 to bring up the Symbol Selector.

Choose the Highway symbol. This time leave it bright red.

Click OK to close the Symbol Selector dialog.

Step 7. Fly Through the Mountains

Flight_path2 takes you through hillier terrain. It's a little more interesting, and more difficult to follow.

Use the Navigate tool to get to a comfortable starting position again.

Click the Fly tool, and practice flying through the hills and valleys of the surface. If you can follow the red path all the way to the end, congratulate yourself!

Step 8. Close ArcScene

Experiment as much as you like with the Fly tool. If you want to save the scene, give it a name of your choice and save it in your 3DDATA\Chapter03\MyData folder.

Exercise 3-4

Create 3D Animated Films

The animation functions in 3D Analyst let you record, play, save, and share films that you create by manipulating data properties and scene perspectives. The animation capabilities are quite numerous, as you'll see; in this exercise, you'll make several short films with datasets that cover a region of the Kentucky River.

Step 1. Start ArcScene and Add Data

Open ArcScene, and click the Add Data button.

Navigate to your 3DDATA\Chapter03\Data folder.

Add cole_doq.jpg, course.shp, raft.lyr, and scole_tin to the scene. (For version 9.1 or 9.2, add raft_v91.lyr instead.)

Step 2. Set Base Heights for the Aerial Photograph

Cole_doq.jpg is a digital orthophoto—an aerial photo that has been georeferenced and corrected for camera tilt and terrain relief.

In the Table of Contents, right-click cole_doq.jpg and choose Properties.

Click the Base Heights tab.

Click "Obtain heights for layer from surface," and choose scole_tin from the dropdown list.

Click OK to close the Layer Properties dialog.

Step 3. Lower the TIN's Drawing Priority, and Turn Off Edge Types

The TIN and the aerial photo are competing for visual space.

In the Table of Contents, double-click scole_tin to bring up its layer properties.

Click the Rendering tab. In the Effects panel, choose a drawing priority of 5 from the dropdown list.

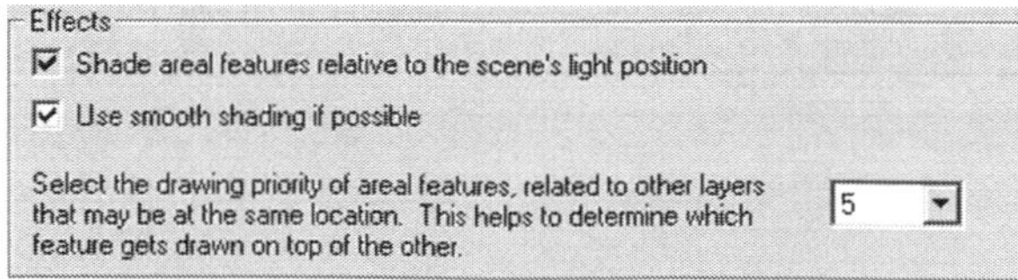

Click Apply.

Now click the Symbology tab. In the Show box, uncheck Edge Types.

Click OK to close the Layer Properties dialog.

The TIN now has a lower drawing priority than the rest of the data, so it doesn't poke up out of the ortho photo.

Step 4. Symbolize the Course Layer

In the Table of Contents, right-click the line symbol under "course" to bring up the Color palette.

Change the color to white.

Your scene should now resemble the graphic below, with scole_tin's drawing priority set to 5, cole_doq.jpg's base heights set to the TIN layer, and course.shp symbolized in white.

Step 5. Create Bookmarks

Click the Full Extent button.

From the Bookmarks menu, click Create. (If you're using an earlier version of ArcScene, you'll find Bookmarks under the View menu.)

Name the bookmark Full Extent.

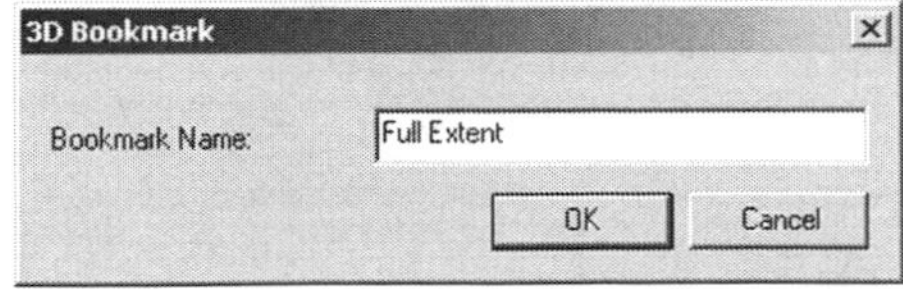

Click OK.

Use the Navigate and Zoom In tools to arrive at the following perspective:

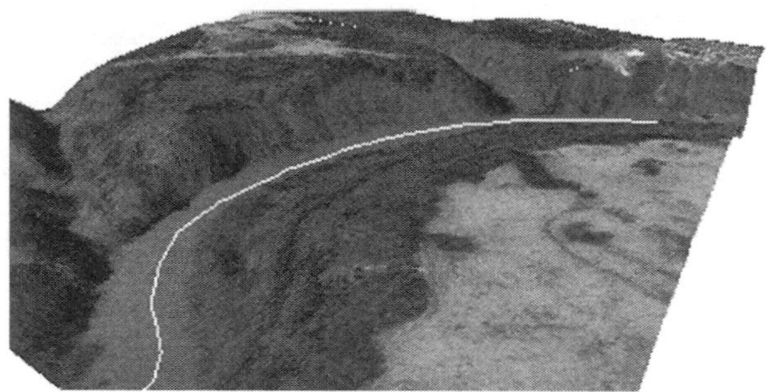

From the Bookmarks menu, create another bookmark. Call it River Bend.

Now zoom to the treatment plant at the northeast corner of the ortho photo.

Create a third bookmark, and call it Treatment Plant.

Finally, navigate around to the north side of the treatment plant. Create a fourth bookmark, and call it Backside.

Step 6. Load the Animation Toolbar

From the View menu, choose Toolbars, then Animation.

The Animation toolbar lets you create films from a series of captured scene settings, layer properties, or camera positions. It works a lot like traditional animation: each camera track is composed of "keyframes" or snapshots of the data. The tracks are stored in the ArcScene document, and can be shared as .asa (ArcScene Animation) files, as .avi files, or as QuickTime movies.

Step 7. Capture Perspective Views to Create Animation

The most basic way to make an animation track is to capture camera positions in a scene and store them as keyframes (snapshots). ArcScene creates the track by interpolating between the keyframes.

Click the Full Extent button again.

On the Animation toolbar, click the Capture View button to make the first keyframe.

Zoom up to the southeast corner of the aerial photo. (Remember that at Full Extent, you're looking at the surface from the southwest.)

Click the Capture View button again.

Now use the Full Extent, Navigate, and Zoom In tools to head over the northwest corner of the photo.

Click the Capture View button to make the third keyframe.

Navigate and zoom to the treatment plant in the northeast corner, and click the Capture View button a fourth time.

Zoom to the center of the photo, and make a fifth keyframe.

Finish up by zooming to the southwest corner, and make a final keyframe.

Step 8. Play Back Your Animation

On the Animation toolbar, click the Open Animation Controls button.

The Animation Controls operate just like a CD or DVD player, with Play, Pause, Stop, and Record buttons.

Move the Animation Controls out of the way of the scene, and click the Play button. 

ArcScene interpolates between the six camera positions you captured, taking a trip from the southeast to the northwest to the northeast to the southwest.

Step 9. Change the Speed of the Animation

Depending on any previous or default Options settings, the animation may be playing too rapidly or too slowly, so you'll change the length of the track.

On the Animation Controls, click the Options button.

In the By duration box, type "7."

Click the Options button again to collapse the dialog box.

Click Play. Now the track takes seven seconds to run.

Step 10. Clear the Animation

From the Animation dropdown menu (not the Animation Controls), choose Clear Animation.

This removes all the animation tracks from the scene. If you'd like to start over and make a new track with the Capture View button, you can clear the animation track at any time.

Step 11. Record Real-time Navigation

In addition to interpolating between camera positions, ArcScene can also record navigation in real time. For example, you could make a film in which the data spins around its axis while you zoom in and out to places of interest. In this step, you'll record navigation with the Fly tool.

If you made some more tracks with the Capture View button, choose Clear Animation from the Animation menu.

Click the Full Extent button.

In the Table of Contents, uncheck cole_doq.jpg, so you can see the TIN underneath.

Click the Fly tool. 

On the Animation Controls, click the Record button. 

Click the Fly tool once in the scene to get oriented, then again to begin the flight. Navigate around the TIN for as long as you like. When you're finished, press the Esc key.

Click the Stop button on the Animation Controls. 

Your flight has been recorded. Click the Play button to watch the animation.

Clear the animation when you've finished watching it. If you like, try recording more navigation with the Fly tool. You can also record sessions with the Navigate tool or with animated rotation (remember, animated rotation is activated from the General tab in the Scene Properties dialog).

Step 12. Use Bookmarks to Make a Camera Track

If necessary, select Clear Animation from the Animation dropdown menu, and click the Full Extent button.

Click the Navigate button to disable the Fly tool, and turn on cole_doq.jpg again in the Table of Contents.

Click the Animation menu, and choose Create Keyframe.

In the Create Animation Keyframe dialog, choose Camera from the Type dropdown list.

Click New to create a new track. The first keyframe is created for you when you make a new track.

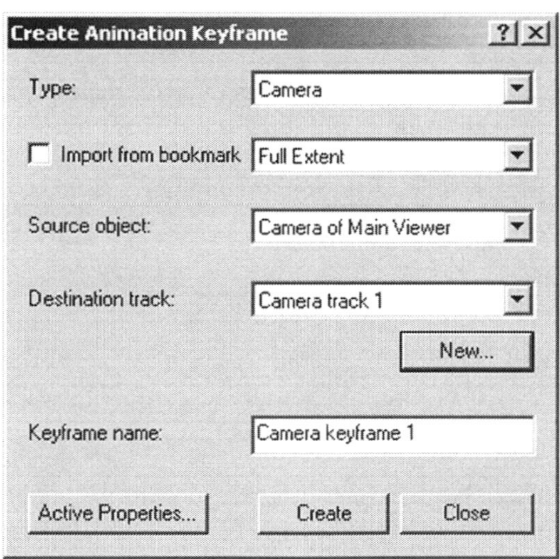

Because Import from Bookmark wasn't checked, ArcScene just used the current camera angle in the scene, which happens to be at full extent.

Now check Import from Bookmark. From the adjacent dropdown menu, choose River Bend.

Click Create.

Now choose Treatment Plant from the dropdown menu, and click Create.

Choose Backside, and click Create.

Choose Full Extent, and click Create one last time.

Click Close.

On the Animation Controls, click the Options button, and type “10” in the By duration box. Click Options again to collapse the dialog.

Click the Play button to run the animation.

Step 13. Add the 3D Effects Toolbar

From the View menu click Toolbars, and check 3D Effects.

The 3D Effects toolbar contains several tools that affect lighting, shading, drawing priority, and transparency. It also has a tool for face culling, which lets you see through areal features by hiding either their back or front views, depending on your observer’s position.

On the 3D Effects toolbar, select cole_doq.jpg from the Layer dropdown list.

Step 14. Change Layer Properties During Animation

From the Animation menu, choose Create Keyframe.

From the Type dropdown list, choose Layer.

From the Source Object list, choose cole_doq.jpg.

Click New, and leave the name “Layer Track 1.”

The Create Animation Keyframe should look like this:

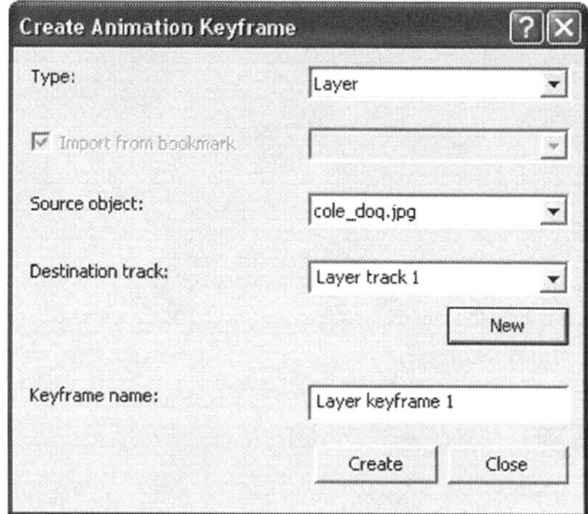

You're going to use the Layer Transparency tool to fade the aerial photo in and out during animation.

Move the dialog out of the way of the scene.

Both scole_tin and cole_doq.jpg should be turned on in the Table of Contents. Also reassure yourself that cole_doq.jpg is selected in the 3D Effects toolbar.

In the Create Keyframe dialog, click Create.

On the 3D Effects toolbar, click the Layer Transparency tool. 

Slide the transparency control all the way from 0 to 100 percent.

In the Create Keyframe dialog, click Create again.

Now slide the transparency control back to 0 percent.

Click Create three more times. This way, the aerial photo will fade up well before the end of the animation track.

Close the dialog.

Step 15. Play Back the Animation

If necessary, click the Open Animation Controls button on the Animation toolbar. 

Press the Play button.

Both tracks play at the same time: the Camera track, which interpolates between the different bookmarks you made, and the Layer track, which fades the aerial photo out and back in again.

Step 16. Open the Animation Manager

The Animation Manager lets you fine-tune your animation in a variety of ways. For example, you can add, delete, enable, and disable tracks; toggle keyframes; adjust the location of keyframes within track play, and change the percentage of total play time that each track fills.

From the Animation menu, choose Animation Manager.

Click the Tracks tab.

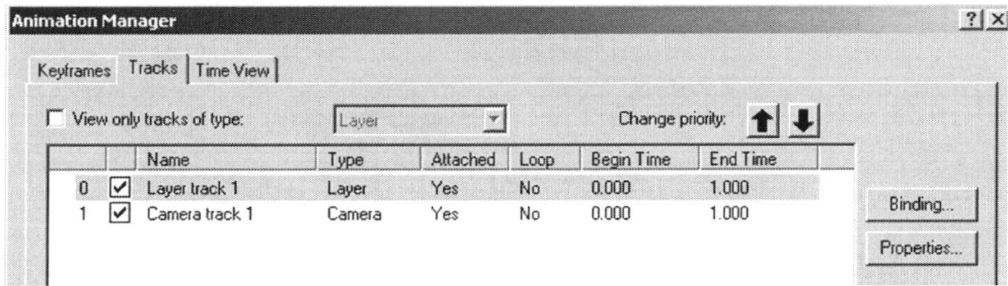

Take a look at the tracks. Both the Camera track and the Layer track are enabled, and they start and end together.

Click the Time View tab.

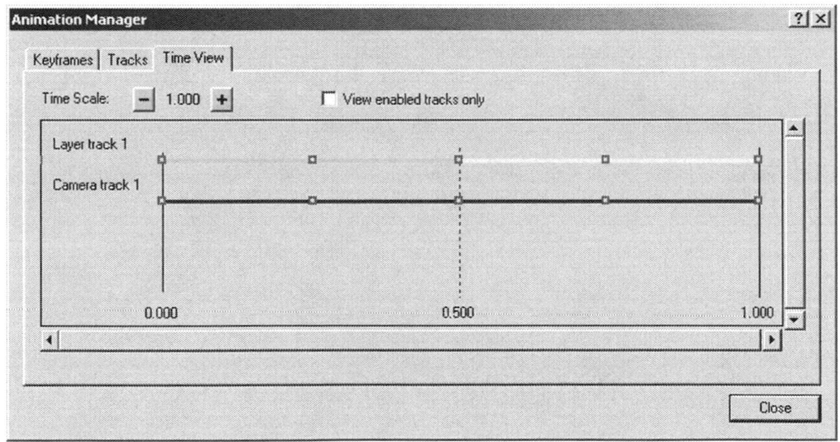

The Camera track and the Layer track each happen to have five keyframes, and in this case they punctuate the track runtime at the same points. The Animation Manager treats keyframes as events that happen along a percentage of the entire play time. (To alter the number of real seconds an animation plays, as you may remember from Step 9, you use the Options button in the Animation Controls.)

Step 17. Preview the Animation at Specific Moments Along the Track

Move the Animation Manager as well as you can away from the scene.

In the Animation Manager, click the cursor anywhere within the timeline. Try to click in negative gray space, rather than on one of the horizontal track lines. A vertical red line appears, and its relative position on the timeline is listed in red at the lower left of the Time View panel.

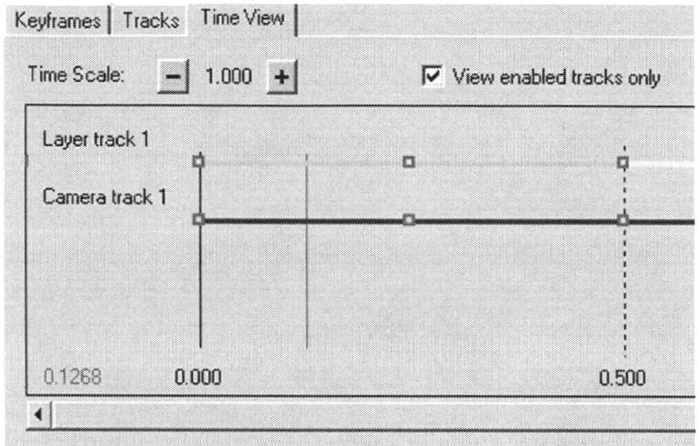

Look at the scene. It shows you what the animation looks like at that point in the track.

Click a few other places along the timeline, and watch the scene.

When you're finished, click anywhere outside the timeline.

Step 18. Set the Layer Track to Play Before the Camera Track

The keyframes along each track are represented by small squares. You can move these wherever you like, and the track will play accordingly.

Using your cursor, grab the squares and move them, one by one, along the Layer track and the Camera track, until their placement resembles the graphic below:

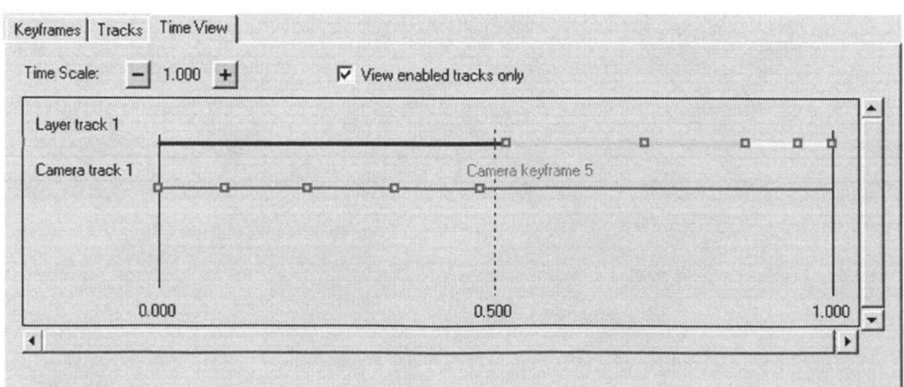

You've set up the animation so that the Camera track runs during the first half of the timeline, and the Layer track runs during the second half.

Click Close.

Step 19. Play the Animation

On the Animation Controls, click the Play button.

This time, the camera makes a path through the bookmarked locations, and then the aerial photo fades out and back in. Notice that each track takes about half the amount of time that it did before.

Feel free to experiment with the Time View settings and the two tracks.

Step 20. Remove the Layer Track and Reset the Camera Track

From the Animation menu, open the Animation Manager. Click the Tracks tab.

Select the Layer track, and click Remove. Only the Camera track is left.

Click the Keyframes tab.

Make sure that "Keyframes of Type" is set to Camera, and the box next to "Distribute time stamps evenly" is checked.

Click Reset Times.

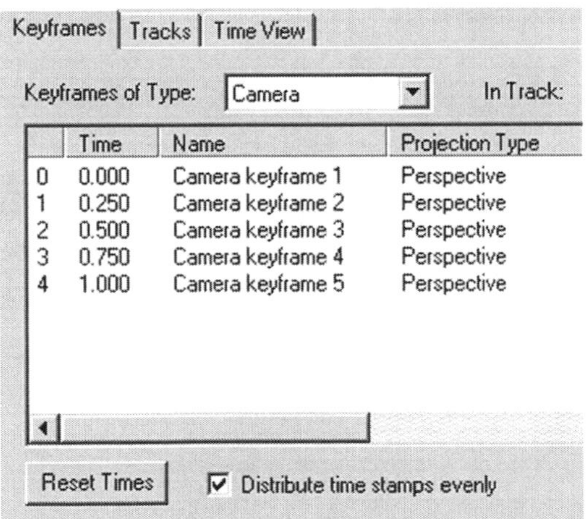

Click the Time View tab, and look at the Camera track. The keyframes are evenly distributed over the timeline again.

Close the Animation Manager, and click Play to run the Camera track again.

Step 21. Clear All Animation

From the Animation menu, choose Clear Animation.

Step 22. Move a Layer Along a Path

The expression "move a layer along a path" may sound somewhat mysterious, but it means just that: you can move the midpoint of any geographic layer along a selected line feature or line graphic in the scene. In this case, you're going to move the raft layer along the course layer.

In the Table of Contents, right-click course, choose Selection, and click Select All.

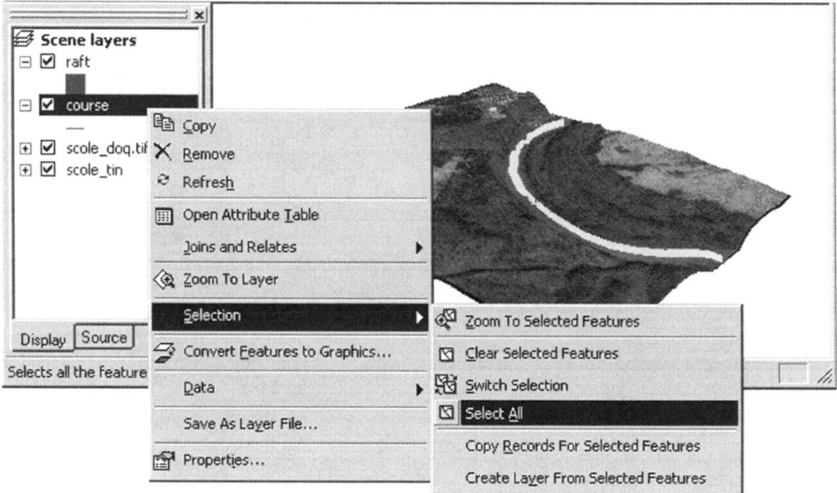

From the Animation menu, choose Move Layer Along Path.

From the Layer dropdown menu, choose raft.

The Path Source is the line feature (course.shp) that you selected in the scene.

In the Vertical Offset box, type "-5." This submerges the raft somewhat so that it looks less like a lumpy brown racecar.

Set the Simplification factor about a quarter of the way along the bar, if it isn't there already. The higher the simplification, the smoother the animation, but the more it taxes your system.

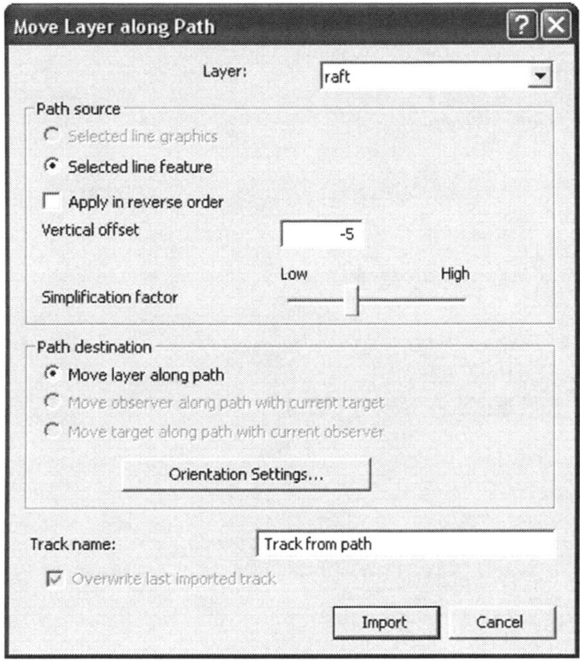

Click Import.

In the Table of Contents, uncheck course. (The line shapefile is still selected, just not visible.)

In the 3D Effects toolbar, select cole_doq.jpg from the dropdown menu. Click the Transparency tool, and set the aerial photo's transparency to about 20 percent.

If the Animation Controls aren't active, click the Animation Controls button on the Animation toolbar.

Click the Options button on the Animation Controls.

In the Duration box, type 10.

Click Options again to close the panel.

Step 23. Play the New Track

Click the Play button.

The raft traces a course along the bend in the Kentucky River. (If you have trouble with this step or Step 25, see the notes at the end of the exercise.)

Step 24. Rename the Track, and Disable It

From the Animation menu, choose Animation Manager.

Click the Tracks tab.

Click on the words "Track from path." A box appears around the highlighted name. Rename the track "raft," and press the Enter key.

Now uncheck the box next to the raft track. You'll use it again later, but for the next step it should be disabled.

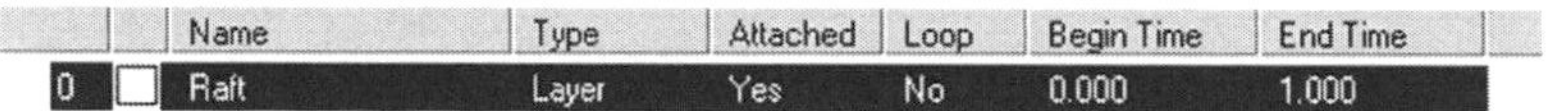

		Name	Type	Attached	Loop	Begin Time	End Time
0		Raft	Layer	Yes	No	0.000	1.000

Close the Animation Manager.

Step 25. Create a Camera Flyby from a Path

You just used the course shapefile to set a path for the raft layer to follow. You can also use that line feature to create a path for a target to move along. In this step, you'll have an observer (the camera) sit in one location and turn to watch the target move along the path.

Zoom and navigate to the outer bank of the bend in the river, like this:

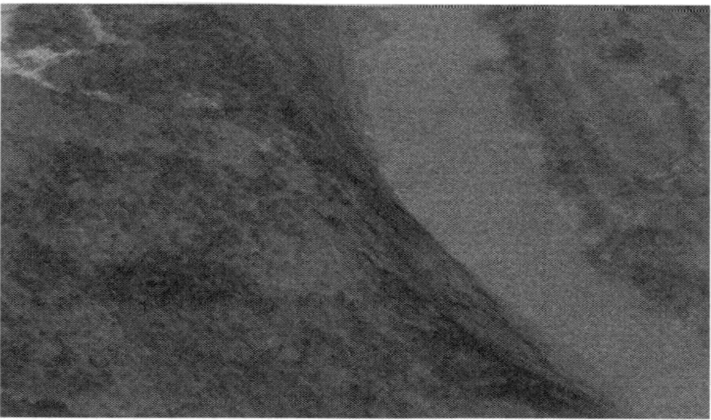

From the Animation menu, choose Create Flyby from Path.

The path source is the course shapefile. It should still be selected, but turned off, in the Table of Contents.

For the path destination, choose the last option, "Move target along path with current observer." This means that the current position of the camera—zoomed into the bank of the river—stays still while the target moves along the path.

Click Import.

Click the Play button on the Animation Controls.

ArcScene moves the (invisible) target along the path that you specified. The camera's location doesn't change, but it does turn to keep the target in view.

Step 26. Play Both Tracks Together

Open the Animation Manager, and click the Tracks tab.

Turn on the raft track. Leave the new Track from Path on as well.

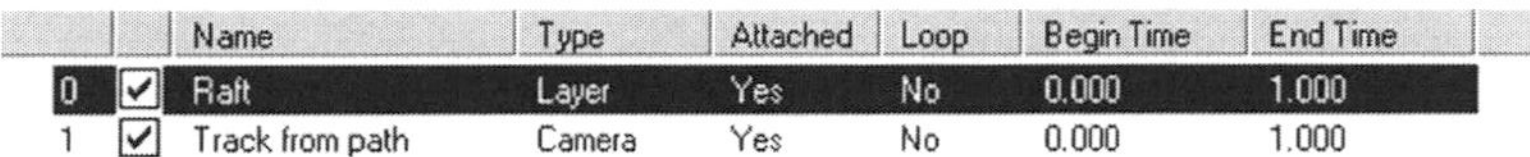

		Name	Type	Attached	Loop	Begin Time	End Time
0	☑	Raft	Layer	Yes	No	0.000	1.000
1	☑	Track from path	Camera	Yes	No	0.000	1.000

Click Close.

Click the Play button.

The tracks play simultaneously: the raft layer travels along the course layer, and the camera observes a target which traverses the same path.

If, like me, you have an older graphics card, you might have trouble with the animation in Steps 22 – 25. If, for example, the raft layer disappears instead of traveling along the path, or the camera flyby misbehaves, try this: Close ArcScene. Open a new empty scene, and add the data that you added at the beginning of the exercise, Steps 1–4. Then skip to Step 22 and try again.

Step 27. Close ArcScene

This exercise only covers a few of ArcScene's animation functions. See the 3D Analyst Help and the 3D Analyst tutorial manual to learn more. In the meantime, feel free to experiment with the Animation Manager.

When you're finished, close ArcScene. If you want to save your changes, name the scene and save it in your 3DDATA\Chapter03\MyData folder.

CHAPTER **4**

ArcGlobe

Like ArcScene, ArcGlobe lets you display and analyze 3D data. It has similar settings for vertical exaggeration, background (sky) color, and sun lighting. There are, however, some major differences in the way that ArcGlobe handles layer display, coordinate space, and your computer's memory usage.

As you remember, ArcScene projects all the data in a scene into the coordinate system of the first data layer that you add. ArcGlobe provides a spherical surface, with default earth-like, upon which to display feature, raster, or terrain data. All data layers are projected into the WGS 1984 Geographic Coordinate System, which references locations in latitude and longitude. As long as your data is in a known coordinate system, ArcGlobe will convert it to GCS WGS 1984 and place it in its proper geodetic location.

In ArcScene, layers can be displayed in 3D by their base heights, if they have them, or they can be draped over other layers that provide those heights. ArcGlobe adds some complexity to this system, categorizing layers as elevation, draped, or floating. The best way to understand ArcGlobe's display rules is to work with them, which we'll do in the next two exercises.

Finally, ArcGlobe has a more sophisticated way of handling memory usage. ArcScene simply loads all data into the available RAM on your computer and uses paging files if necessary. ArcGlobe lets you set up memory caches based on data types or for individual layers. We'll look at caching in Exercise 4-2.

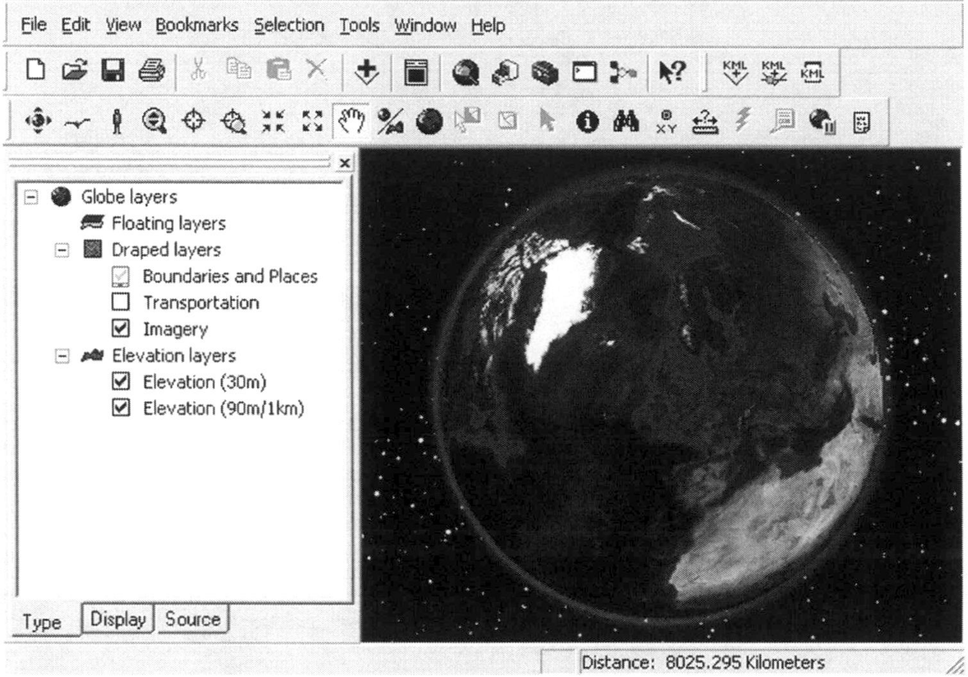

The default ArcGlobe interface for 3D Analyst (version 9.3)

Exercise 4-1

Understanding ArcGlobe

In this exercise, you'll take a look at ArcGlobe's default layers, check out ArcGlobe's display properties, and use the more basic navigation tools.

Step 1. Start ArcGlobe

Double-click the ArcGlobe icon on your desktop. If you didn't make an icon in Chapter 1, click the Start menu, point to Programs, point to ArcGIS, and click ArcGlobe.

Step 2. Load Toolbars

In ArcGlobe, the 3D Analyst extension should be loaded by default. Check this by clicking on the Tools menu, then Extensions. The box next to 3D Analyst should be checked.

From the View menu, choose Toolbars, then check the boxes next to "Standard" and "Tools." This loads the toolbars that you'll use most often.

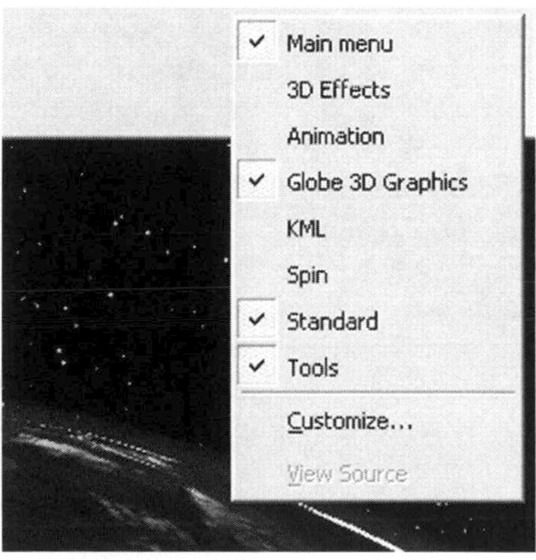

Step 3. Use Some Key Navigation Tools

On the Tools toolbar, click the Full Extent button, 

and then the Navigate tool. 

The Navigate tool should be very familiar to you by now. It works almost the same way here as it does in ArcScene. Holding down the left mouse button rolls the globe up, down, and around, keeping the center of the spheroid as its point of reference. (If you have version 9.3, the background of stars rotates also, proving poor old Copernicus wrong after all.) Holding down the right mouse button lets you zoom in and out, as does scrolling with the center mouse button.

Experiment with the Navigate tool for a while. When you have the hang of it, click the Full Extent button.

Now, click the Navigation Mode button, again on the Tools toolbar.

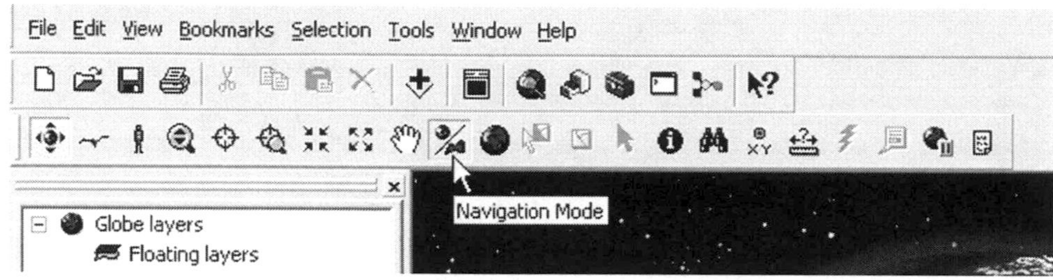

This is a toggle button that switches between Globe mode and Surface mode. When the button looks depressed, you're in Globe mode. That's the default, and it's what you've been working with so far.

When the button looks flush with the rest of the toolbar, you're in Surface mode. Surface mode seems pretty annoying at full extent; navigating isn't very easy. Zoom in to a close location, though—say the middle of the United States—and Surface mode starts to make more sense. While Globe mode places the center of movement at the center of the earth, Surface mode places the center of the view on the surface, wherever you have zoomed to. This way, when you navigate the camera acts more like ArcScene, allowing you to tilt away from the globe's surface and "fly in."

Holding down both mouse buttons allows you to pan, but only in Surface mode. Panning is disabled in Globe mode; in fact holding down both mouse buttons automatically switches you to Surface mode.

After navigating around in both modes a bit, you'll be happy to discover the Center on Target and Zoom to Target buttons, on the Tools toolbar.

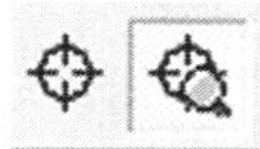

These tools work the same way in ArcGlobe as they do in ArcScene, but they are even more useful here, since the area covered is much larger. You will spend less time struggling with the mouse and the scroll wheel if you remember that these buttons are your friends.

Step 4. Toggle Between Normal and Draft Mode

Zoom in closely to a land location and click the Draft Mode button, located on the Tools toolbar.

The button turns orange, and the resolution of the image becomes coarser. Draft mode lets you render data and imagery faster while navigating.

A view of ArcGlobe's default layers in both normal and draft modes

The settings of Draft mode can be controlled in the Level of Detail panel, located in the Tools menu under Options. You can control the number of features shown and the level of detail used for elevation and raster layers. You can also disable the drawing of 3D marker symbols and 3D object textures.

Draft Mode
Activate
When the draft mode is active:
Reduce detail of image and elevation data
Disable 3D marker symbols
Disable 3D object textures
Disable vector line densification
Draw only one feature for every:
4 vector features

One of the more difficult concepts to grasp in ArcGlobe is its categorization of layers as they appear on the globe's surface. Your knowledge of base height settings in ArcScene will help you in some cases and hinder you in others. Let's take a look at ArcGlobe's default layers.

Make sure that the Type tab is selected at the bottom of the Table of Contents.

There are really four types of surface used in ArcGlobe: the bare globe surface itself, elevation layers, draped layers, and floating layers. Elevation layers give shape to the globe surface. They are often elevation rasters, TINs (as of version 9.3), or terrains, which you'll learn about in Chapter 8. These elevation layers behave the same way in ArcGlobe as they do in ArcScene, providing their own base heights.

- Globe layers
  - Floating layers
  - Draped layers
    - [x] Boundaries and Places
    - [ ] Transportation
    - [x] Imagery
  - Elevation layers
    - [x] Elevation (30m)
    - [x] Elevation (90m/1km)

Draped layers are usually raster images or 2D vector features. As in ArcScene, they use elevation layers to get their base heights and so are "draped" on those surfaces.

Floating layers relate to the bare globe surface, using a simple offset to float either above or below it. Floating layers can be rasters or vector features, and even TINs and terrains, although you usually want layers to float that don't represent topographic surfaces. Clouds, ozone levels, or rasters showing the spread of disease would make more likely floating layers.

Step 5. Examine ArcGlobe's Default Elevation Layers

In version 9.2, there were no default elevation layers giving "bumps" to the globe surface. In version 9.3, there are two elevation layers loaded: 30-meter resolution for the US, and 90-meter resolution for the rest of the world. These layers are being streamed from ESRI's online servers, so if you have no Internet connection, you won't see them (unless ArcGlobe has had online access to them before and has built a cache for the layers).

In the Table of Contents, turn off both elevation layers.

From the View menu, choose Globe Properties. Click the General tab.

Set the Globe display units to Meters, and check the boxes to show Latitude, Longitude, and Elevation.

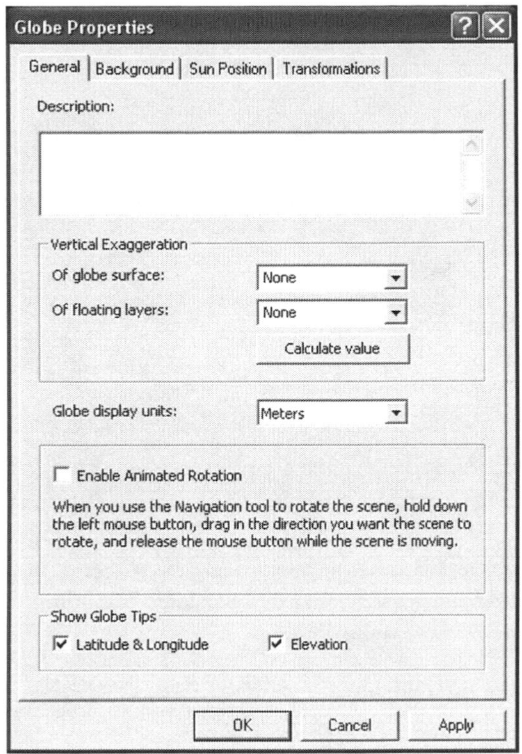

Click OK to close the dialog.

Pan around to various locations within the United States, each time letting the mouse cursor rest for a moment. The tooltip should show latitude and longitude in degrees minutes and seconds, and elevation in meters. However, since the elevation layers are turned off in the Table of Contents, the elevation registers 0.00 each time. You can think of the bare spherical surface provided by ArcGlobe as sea level, which is zero elevation.

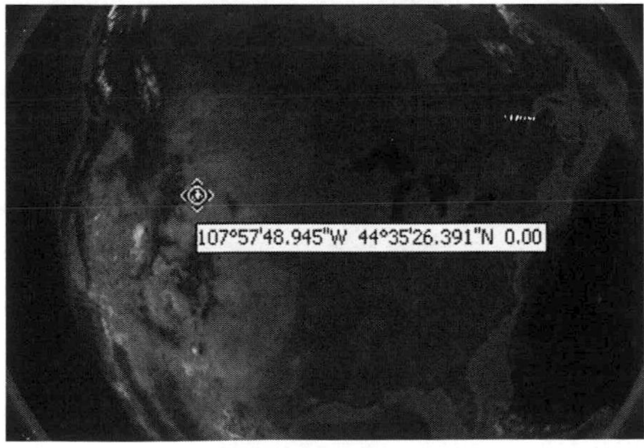

Turn the 30-meter elevation layer on in the Table of Contents.

Now when you pan across the United States, the tooltip shows the elevation. (The units are meters, because that's how you set the Globe Display units on the General tab of the Globe Properties dialog.)

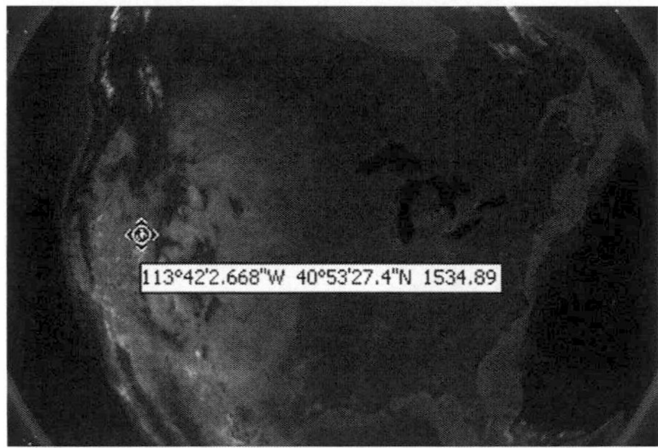

If you pan outside the United States, however, the elevation registers at 0.00 again, because you still have the 90-meter world elevation layer turned off in the Table of Contents. Turn it on, and the tooltip will show the elevation values provided by that layer.

Make sure that the 90-meter elevation layer is on in the Table of Contents, and click the Navigate tool.

Zoom in to a mountainous area outside the United States, such as the Himalayas, and make sure that the Navigation Mode button is set to Surface (remember, it looks flush with the rest of the toolbar when Surface mode is active).

Turn off all of the layers in the Table of Contents except for 90-meter elevation layer.

The elevation layer itself is invisible, but it gives shape to the globe surface. It also provides elevation values for the tooltip.

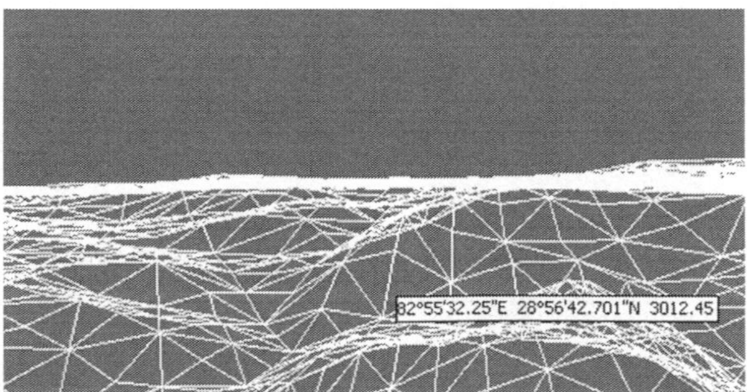

Now, turn off the 90-meter elevation layer in the Table of Contents. Note that the spherical, sea-level surface of ArcGlobe is all that remains, and the elevation value in the tooltip returns to 0.00.

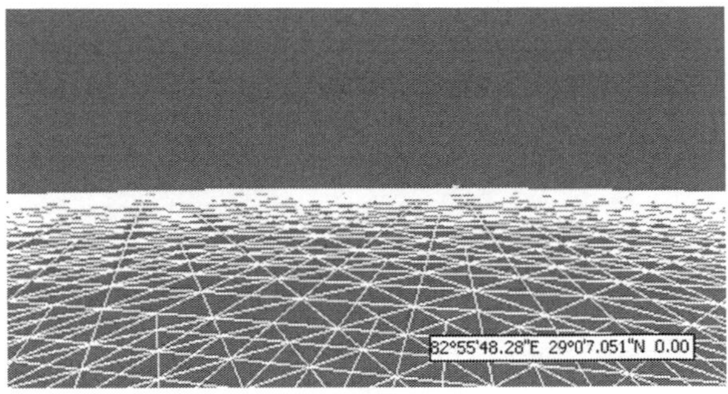

When a layer is categorized as elevation in the Table of Contents, it is invisible. Its job is to give shape to the globe's surface and the layers draped upon it. Furthermore, elevation layers behave in ArcGlobe's Table of Contents as they do in ArcMap; they draw in the order that they are listed, from top to bottom. Although the ArcGlobe Help suggests that the elevation layer with the highest resolution is chosen to give shape to the surface, my testing with the 90-meter and 30-meter elevation layers shows otherwise. So when you work with your own higher-resolution elevation datasets, make sure that they're on the top of the stack of elevation layers in the Table of Contents.

Step 6. Examine ArcGlobe's Default Draped Layers

In ArcGlobe version 9.3, there are three draped layers: Imagery, Boundaries and Places, and Transportation.

Zoom to full extent, and make sure that the Navigation tool is selected.

In the Table of Contents, turn on all three draped layers.

You'll notice that at this extent, the check boxes next to the Boundaries and Transportation layers are grayed out. In order to view the layers, you need to be zoomed in closer.

First, though, right-click the Imagery layer in the Table of Contents, and choose Properties.

Click the Globe General tab. Notice that in the Distance Range panel the layer is set to display at all distances. This is a pretty cool layer; if you read the description, you'll see that it incorporates several imagery datasets at different resolutions so that it can be viewed seamlessly from any distance.

Now click the Source tab.

Like the two elevation layers, the data sources for the three draped layers are also being streamed from ESRI's online servers.

Click the Elevation tab. Note that the Imagery layer is being draped "on the globe surface." This is a little misleading, because the layer is really being draped on whatever elevation surface is available—in this case, the 90- and 30-meter elevation layers. If you were to turn those layers off, then the Imagery layer would truly be draped on the bare globe surface.

Close the Layer Properties dialog for the Imagery layer.

In the Table of Contents, right-click Boundaries and Places and choose Properties.

Click the Globe General tab, and take a look at the Distance Range panel. This layer is set to display only if you're zoomed in past 1,800,000. One million eight hundred thousand what, you ask? Well, remember in Step 5 that you set the Globe display units to Meters. That affects not only the elevation of the features shown on the surface but also the "viewport distance" between the observer and the globe's surface. If you set the display units back to Kilometers, the number in the Distance Range panel would change to 1,800.

Zoom in far enough to render the Boundaries and Places layer visible. The program will attempt to resize the labels to an appropriate scale, just as in ArcMap.

After familiarizing yourself a bit with the information in the General, Source, and Elevation tabs, open the Layer Properties for the Transportation layer. What is the distance range for its visibility? ____________

ArcGlobe doesn't come with default floating layers, so in the next exercise we'll add our own.

When you've finished exploring the properties of the draped and elevation layers, close ArcGlobe. Say No when prompted to save changes.

Exercise 4-2

Explore ArcGlobe's Options, Add Data, and Redefine Layer Types

Step 1. Examine ArcGlobe's General Application Options

Start a fresh session of ArcGlobe. From the Tools menu, choose Options.

Click the General tab.

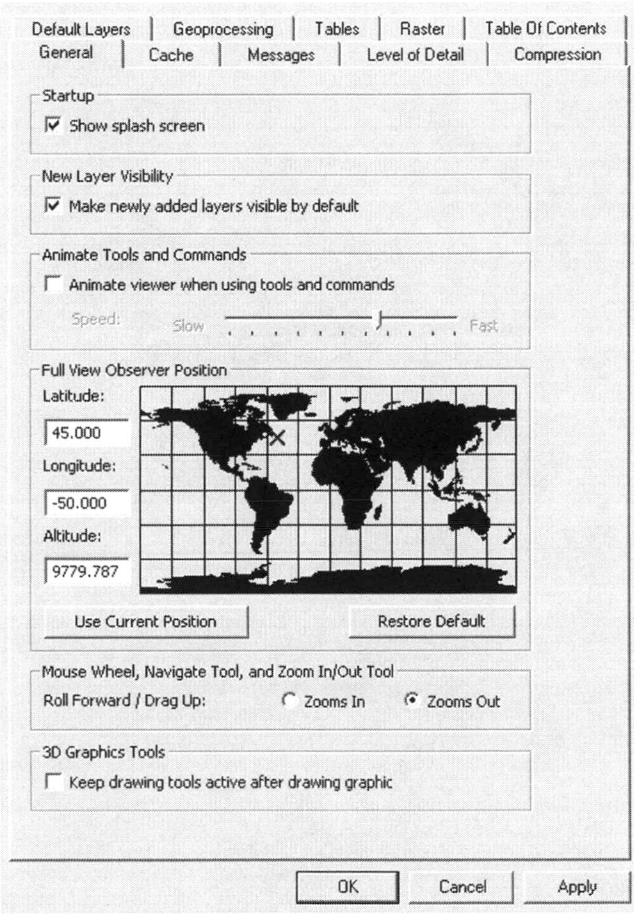

There are some interesting items on the General tab. Many of these options won't take effect until you restart ArcGlobe, but a few, such as the Full View Observer Position, can be applied right away. This option lets you decide what position you'd like the observer to take when ArcGlobe starts up. You can use the little red "X" to set the position, or type in longitude, latitude, and altitude coordinates.

In the Full View Observer Position panel, leave the latitude setting at 45.000 degrees, but change the longitude setting from –50.000 to 90.000 by typing the value in the box.

Notice that the red "X" changes its position. Click OK in the Options dialog.

In ArcGlobe, click the Full Extent button. The view centers on the northern part of western China, between the Kazakhstan and Mongolian borders.

From the Tools menu, choose Options again, and in the General tab click Restore Default. The red X moves back to the default location.

Click Apply, but don't click OK or you'll close the Options dialog.

One of the nicest additions to ArcGlobe, ArcMap, and ArcScene at version 9.3 is the ability to reverse the behavior of the scroll wheel on the mouse. If you use Google Maps and Earth, you know that you have to readjust your thinking to zoom in and out with the scroll wheel every time you switch between Google and ESRI products. For Google, scrolling the mouse wheel toward you means zoom out; for ESRI it means zoom in. I personally think ESRI's way is more intuitive, but in the "Mouse Wheel, Navigate Tool, and Zoom In/Out Tool" panel, they've given us the option of switching to Google's mouse style.

You don't need to change the mouse wheel behavior, but remember where this option is in case you want to come back to it.

Step 2. Examine Cache Settings in ArcGlobe

Click the Cache tab on the Options dialog.

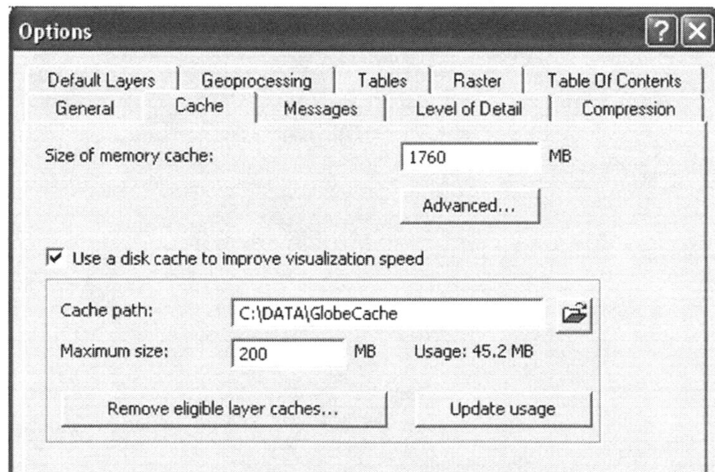

In order to seamlessly display data that is either very detailed (textured multipatch features, for example) or covers a large area (satellite images mosaicked together) on a revolving globe, some rather sophisticated data tiling, storage, and memory management has to go on behind the scenes. At version 9.3, ArcGlobe has added a number of ways to tailor these settings based on the types of data you're using, the age of your graphics card, how much RAM you have, and how much disk space.

The ArcGlobe Help files contain detailed explanations of cache management, disk cache formats, group layer caching, disconnected caching, full, partial, and on demand caching, cache creation and deletion at the layer or application level, and many more other specifications that you'll be interested in if you use large datasets or find your display slowing down, but ArcGlobe caching really breaks down to two major types: Memory and Disk. Let's look at the Memory cache first.

Memory caching just refers to the amount of RAM that ArcGlobe uses. I have mine set to 1760 MB. The example in Exercise Nine of the 3D Analyst Tutorial (version 9.3) asks you to set it at 500 MB. It really is up to you; everyone's needs are different. The only stipulations are that you can't assign more memory cache than you have RAM in your computer.

On the Cache tab, click the Advanced button.

Advanced Memory Cache Settings

Memory cache size : 1760 Mb

Set usage limits per type

	Limit (Mb)	Current (Mb)
Elevation :	100	2.8
Point features :	300	0.0
Line features :	300	0.0
Text :	100	0.0
3D objects (geometry) :	300	0.0
3D objects (texture) :	200	0.0
Images/rasterized :	460	120.0
Total texture limit :	660	

In systems with AGP graphics cards, it should be kept below the AGP aperture size.

OK Cancel

You can set the amount of RAM that each data type in ArcGlobe will use. If you tend to work mostly with DEMs, terrains, and TINs, you might want to allocate more memory to that data type. If you do close-up work designing building facades with textured multipatch 3D objects, you might want to focus your resources there. Again, it's up to you. The nice thing is that ArcGlobe tells you here how much RAM you're currently using for each data type, so you can educate yourself pretty quickly about what you're going to need in the future.

If you don't assign enough RAM to a given data type, ArcGlobe won't balance the numbers out for you. You'll see reduced performance in that case, so it's best to pay attention to these individual values.

Click OK to close the Advanced Memory Cache Settings tab, and look again at the basic Cache tab.

The second type of ArcGlobe caching is Disk caching. Instead of using RAM, ArcGlobe creates additional temporary files for specific layers used in an ArcGlobe document. You can set the disk cache for each layer in its Layer Properties dialog, but you must also decide where you'd like these files to generally reside, and set that path in the Disk Cache panel.

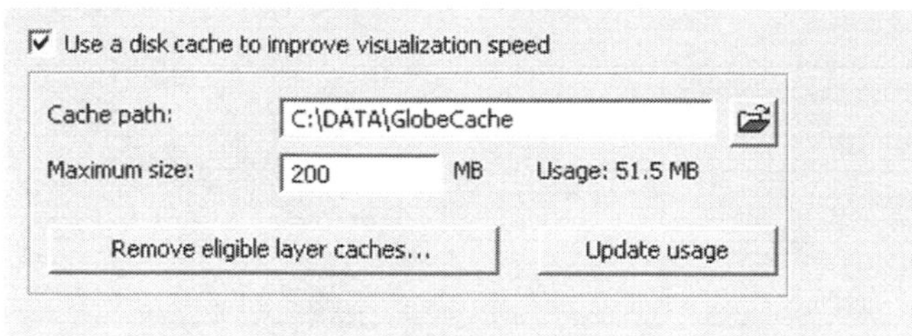

I have told ArcGlobe that Yes, I want to use a disk cache to improve visualization speed, and I have set the path to C:\DATA. ArcGlobe created the folder GlobeCache on its own.

As with Memory caching, ArcGlobe tells you in this panel how much space you're using currently, so you can take better guesses about how much you should allocate in the future. (You have to click Update Usage to refresh the current value.)

I currently have 200 MB allocated for the Disk cache, and ArcGlobe tells me that I'm using about 51 MB.

If I use Windows Explorer or My Computer to browse to C:\DATA\GlobeCache, the files that I've been using show up like this:

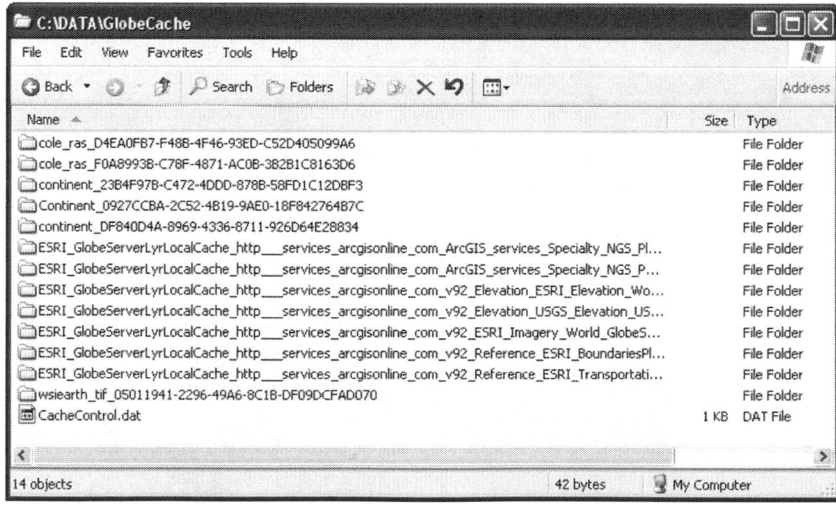

This list tells me what layers in ArcGlobe I've been creating a Disk cache for. Some of them are layers that are currently in the ArcGlobe document, and some are layers that are no longer in use. You can specify how you want the Disk cache handled in the Layer Properties dialog for each layer in ArcGlobe.

Click OK to close the Options dialog.

Step 3. Add a Rasterized Layer to ArcGlobe, Apply a Coordinate System Transformation, and Change Its Transparency Level

Remember that in the last exercise you learned the difference between elevation layers and draped layers. In this step, we're going to add a draped layer, and later we'll redefine it as floating.

In ArcGlobe, make sure that the tab at the bottom of the Table of Contents is set to Type. Also make sure that the 30-meter elevation layer is checked on.

At the top of the Table of Contents, double-click "Globe Layers" to open the Globe Properties dialog. (Remember that this is where you set a number of options that apply to the current document; don't confuse this with the Tools menu/Options dialog, where you set options that apply every time you open the application.)

On the Globe Properties dialog, click the General tab.

Set the Vertical Exaggeration of the Globe Surface to 2, and click OK to close the dialog.

Click the Add Data button.

Navigate to the 3DDATA\Chapter04\Data folder, and add elk_park_bnd.shp.

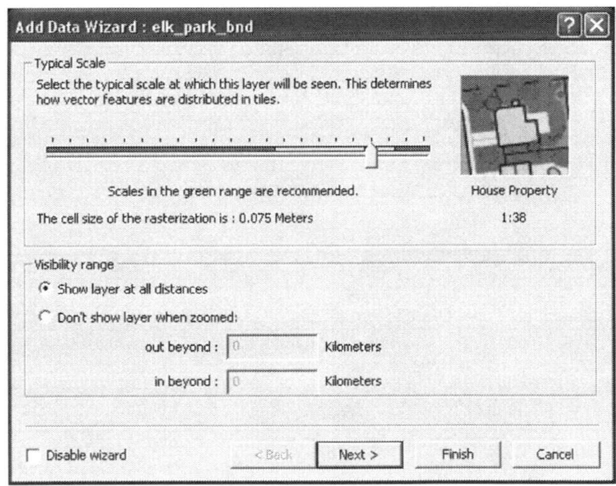

The Add Data Wizard asks you if you want to set a visibility scale. In this case, we want to show the layer at all distances.

One thing to notice is the message about the cell size of the rasterization. ArcGlobe is calculating this from the extent of the layer. I'll explain "rasterization" in a few moments.

Click Next.

The wizard tells you that it will convert point symbols to real-world units, and asks you what real-world resolution and units you'd like to use. Accept the suggestion of 5 Meters, and click Finish.

The Geographic Coordinate Systems Warning appears, telling you that elk_park_bnd uses the NAD 1983 datum, while ArcGlobe uses WGS 1984.

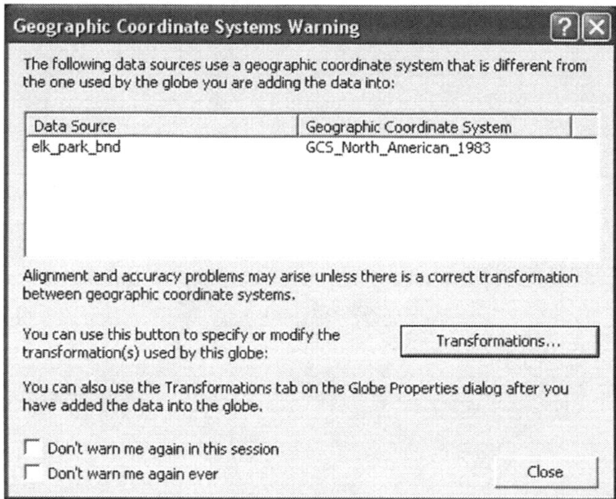

Click the Transformations button.

The Geographic Coordinate System Transformations dialog offers to convert NAD 1983 to WGS 1984, using one of the transformation formulas listed in the dropdown box.

If you right-click in the dropdown box and click "What's this?" you'll find a nice explanation of how this process works. In fact, the example they use states that to convert NAD 1983 to WGS 1984 within the contiguous United States, you can generally use the NAD_1983_to_WGS_1984_5 transformation.

So, from the dropdown list, choose NAD_1983_to_WGS_1984_5.

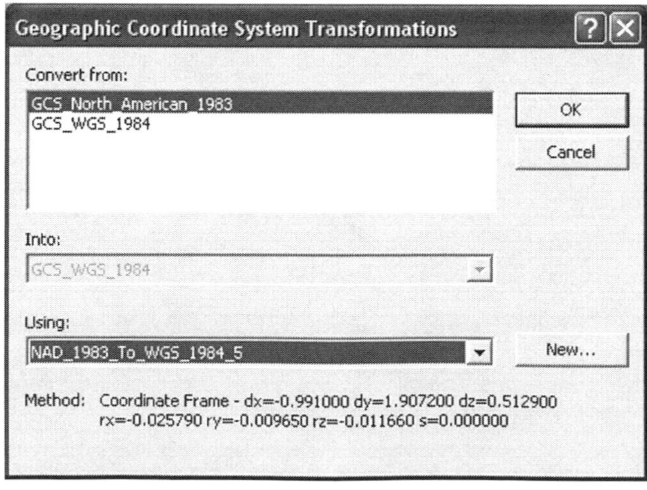

Click OK.

Click Close when the Warning dialog comes up again.

Finally, elk_park_bnd.shp is added to ArcGlobe. It is added as a Draped layer by default—take note of its place in the Table of Contents.

In the Table of Contents, drag elk_park_bnd so that it sits just above the Imagery layer.

Right-click elk_park_bnd, and select Zoom to Layer.

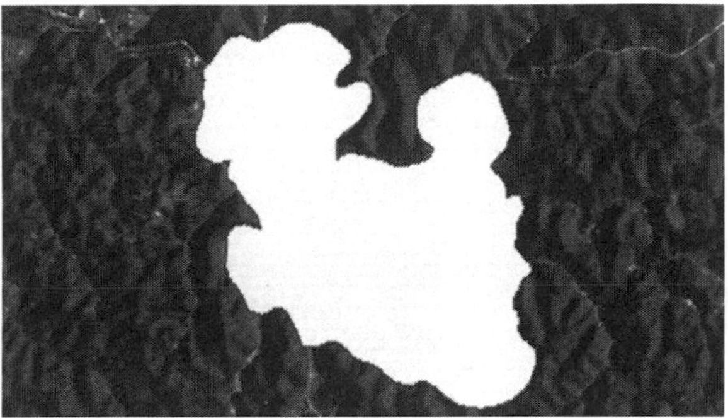

Make sure that the Navigate button is selected, and that the Navigation Mode button is in Surface mode (remember, it looks flush with the rest of the toolbar in Surface mode).

Zoom in, tilt the camera around to get a good idea of the lay of the land, and notice that even though elk_park_bnd is a polygon feature, it drapes nicely over the terrain. Normally, we wouldn't expect this—as you learned in Chapter 2, only the vertices of polygons hold z values, so polygons don't usually drape well over terrain at all.

ArcGlobe, however, "rasterizes" 2D points, lines, and polygons. In other words, it turns this polygon into an image, and then drapes it over the elevation surface. In fact, if you look closely at the edges of elk_park_bnd, you'll see the familiar "jaggy" edges of the cells making up the raster. It isn't too bad, though, because in the Add Data Wizard you set it to a resolution of 5 meters, which is pretty fine.

If you were to add 3D vector points or lines, ArcGlobe would give you the option of rendering them as vector features, since they provide their own height sources. However, rasterizing vector layers makes sense in ArcGlobe for a number of reasons; it is easier for ArcGlobe to maintain the symbology of a layer file if it's rasterized, and rasters can be displayed much faster than vector features. And, as you've seen, polygons display in a reasonable manner over terrain if they're treated as rasters.

Let's make this even more obvious by changing the transparency of elk_park_bnd.

In the Table of Contents, double-click elk_park_bnd to bring up its Layer Properties.

On the Symbology tab, choose a fairly light, bright color, such as off-white.

Click the Display tab. In the box next to "Transparent," set the value to 45%.

Click OK.

Now elk_park_bnd has a transparency that lets you see just how well it drapes over the 30-meter elevation layer in ArcGlobe.

Step 4. Redefine Elk_park_bnd as a Floating Layer

In the Table of Contents, right-click elk_park_bnd, select "Redefine Layer," and then select "Redefine layer as floating."

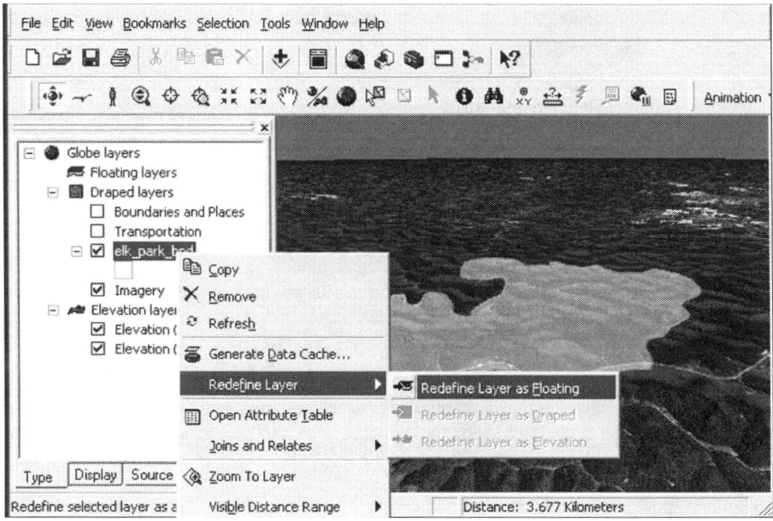

The layer disappears.

In the Table of Contents, right-click elk_park_bnd again and select Zoom to Layer.

Notice that the Navigation mode automatically switches from Surface to Global. The layer is still invisible, though.

Turn off the Imagery layer, and you'll see that elk_park_bnd is floating, but floating below it. In fact, it's floating below both elevation layers, sitting on the bare globe surface. By definition, a floating layer floats on top of a surface of constant elevation, with an offset.

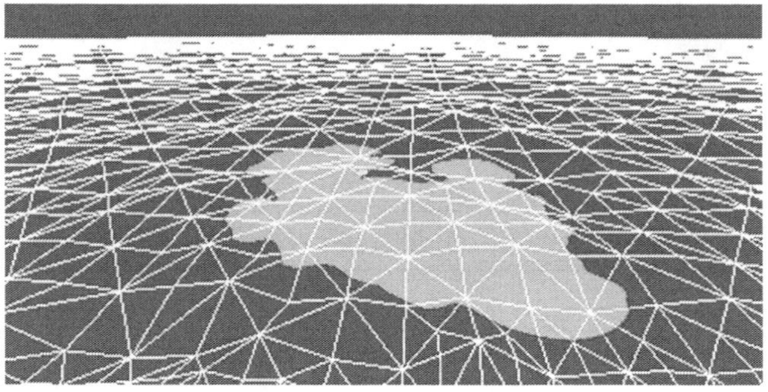

Right_click elk_park_bnd in the Table of Contents, and choose Properties.

On the Elevation tab, note that the Type panel tells you that the layer is floating.

In the Offset panel, type "1000" in the box to make elk_park_bnd float above the surface. Click OK.

Turn the Imagery layer back on. Now elk_park_bnd floats above the terrain. Because it floats on a surface of constant elevation, it appears flat. In some cases, this is what you want. In some cases, you want a layer to conform to terrain, or to its own values not associated with terrain (as in clouds or ozone levels), but to also float above the globe.

Open elk_park_bnd's Layer Properties again. This time set it to float independently of the surface but draped on the 30-meter elevation layer, by choosing it in the dropdown list.

Click OK.

Now elk_park_bnd floats above the surface but also conforms to the shape of the elevation values beneath it.

Earlier in Step 3, you set the vertical exaggeration of the globe surface (or the elevation layers upon it) to 2.

Open the Globe Properties dialog by double-clicking Globe Layers at the top of the Table of Contents.

Give elk_park_bnd the same lift that the elevation layers have by setting the vertical exaggeration of floating layers to 2.

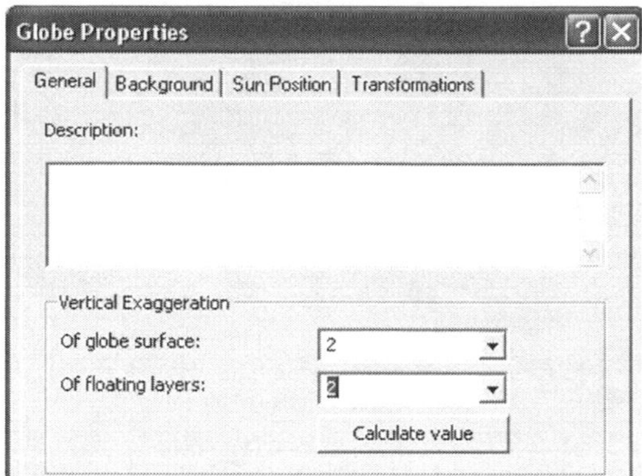

Why not just define elk_park_bnd as a draped layer and give it an offset, you ask? The answer is that when a layer is draped, the option for giving it an offset in unavailable. Therefore, if you want a layer to be offset from the surface for any reason, you have to define it as floating.

Let's try one last trick with elk_park_bnd.

Open its Layer Properties again, and click the Globe Extrusion tab.

Check the box in order to "Extrude features in a layer." In the expression box, type "500."

Click OK to close the dialog.

You can create some pretty cool effects with floating layers in ArcGlobe. Experiment with as many of the layer and globe properties as you like. When finished, exit ArcGlobe without saving changes.

CHAPTER 5

Google Earth

Google Earth is an online virtual globe program that offers anyone with a computer and an Internet connection the ability to zoom to any part of the earth and view aerial photography, terrain surfaces, political boundaries, street names, and places of interest. Google Earth has also revolutionized the web experience by creating a forum where users can create their own 3D models and share them in Google Earth through KML, or Keyhole Markup Language. KML was invented by Keyhole, Inc., along with the original interface, called Earth Viewer. KML uses a tag-based structure similar to HTML and XML that can be created or altered in any text editor. In this tutorial, though, we'll focus on getting around in Google Earth using the normal interface, and importing KML files from ArcGIS into Google Earth.

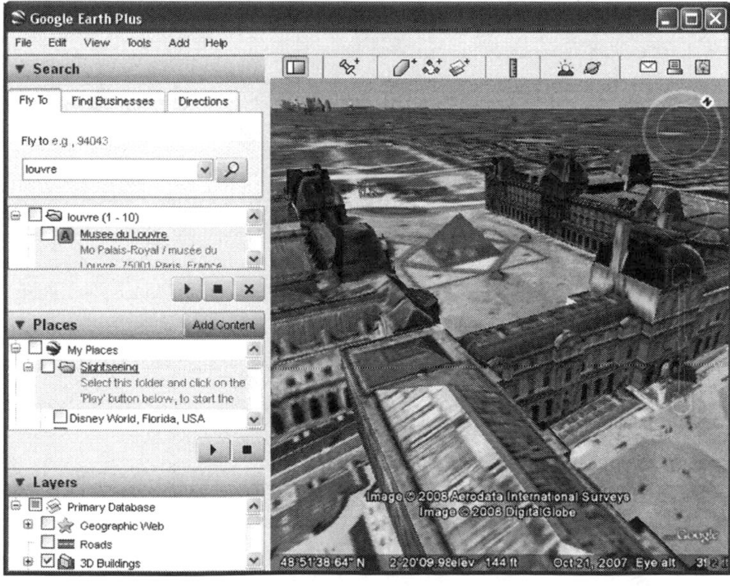

The Google Earth interface

To do the exercises in this book, you need the free version of Google Earth. I use Google Earth Plus, but it is being phased out, leaving only the free version and Google Earth Pro, which costs $400 a year. Because download and installation instructions change quickly on the web, I'm simply going to point you to:

http://earth.google.com

where, as of mid 2009, you can find all the tools necessary to get you up and running. In the next few exercises, you'll learn the ins and outs of Google Earth, and how to use it to display your 3D GIS data.

Exercise 5-1

Navigating Google Earth's Interface, and the Planet

The Google Earth interface is fairly simple and intuitive for most people, but if you're used to ArcGIS, Google Earth can take some getting used to.

Step 1. Start Google Earth

When you install Google Earth, it creates a shortcut on your desktop that you can double-click to start the program. (If you don't see a shortcut, go to Start > Programs > Google Earth and start it from there.

Take a look at the interface. The main components of the Google Earth are the 3D viewer and three side panels: the Search panel, the Places panel, and the Layers panel. You can toggle the visibility of the three side panels by clicking the arrows next to their names. In the following graphic, the Search panel and the Layers panel are visible, and the Places panel is not. (If you don't see the side panels, click the View menu and check Sidebar.)

The Search Panel is probably the most familiar aspect of the application. It works just as it does in Google Maps, Yahoo Maps, or MapQuest. The Directions tab lets you specify starting and destination points for driving directions. On the Fly To tab, you can type in a latitude-longitude pair in any part of the world, or an address, city, state, country, or zip code in some parts of the world, and Google Earth will fly to that location. The Find Businesses tab looks for directory listings and allows you to use partial names instead of exact addresses. If the Fly To tab isn't able to process a specific address search request, the Find Businesses engine will take over and treat it as a directory listing search.

Step 2. Search for Addresses

In the Search panel, click the Fly To tab. Type in a full address, with street number, name, state and city, and click Search. I used "10 Market St, San Francisco, CA."

Click the Search button, or just press Enter.

Google Earth flies to the location.

On the Fly To tab, type in simply "louvre."

Since you're not giving the full address, Google Earth treats it as a directory lookup request. It flies to the general location of the Louvre in Paris and provides a number of possible addresses for you to choose from.

Step 3. Look at the Layers Panel

The Layers panel houses geographic data of many types. The Primary Database is the top folder. You can turn off all the layers at once by clicking the highlighted box next to its name. Underneath that, the folders within the Primary Database each contain subfolders, allowing you to specify which individual layers you want to see. The Terrain layer performs the same function that Elevation layers perform in ArcGlobe; that is, it gives shape to the bare globe surface. For our purposes, the Terrain layer and the 3D Buildings layer are good to keep toggled on.

The only layer that is always visible is the overall imagery layer. Whether you turn the Terrain layer on or off, the imagery is still there. Google gets the imagery from various sources, updated at various times. The imagery is static, indexed and mosaicked in a tiling system similar to ArcGlobe's tiling system. The copyright credits for the particular imagery dataset that you're seeing are always front and center in the view, and change as you move around the earth, or zoom in and out. Spend some time getting to know the contents of the Layers tab.

Step 4. Check Out the Places Tab

The difference between Layers and Places is not so much a matter of data type but of data source. Loosely, it makes sense to think of Places as unique points of interest and Layers as features that exist worldwide, such as roads and political borders; so strictly speaking, we should not be confused by the fact that the Layers tab contains a folder called Places of Interest because these places of interest—everything from parks and recreation to ATMs—are recurring features spread out over large areas. The main difference between Places and Layers is that Layers are streamed from Google's servers, and you can't alter them. Items in the Places tab are simply KML documents that reside on your computer, either in the Google folder or added by you. You can right-click an entry in the Places tab and choose "Copy," then "Paste" into Notepad, and you will see the KML code. You can't do that with the entries in the Layers tab.

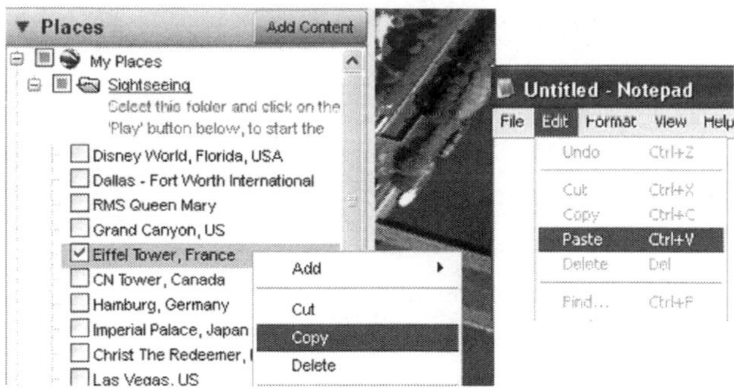

```
<?xml version="1.0" encoding="UTF-8"?>
<kml xmlns="http://earth.google.com/kml/2.2">
<Document>
	<name>KmlFile</name>
	<Style id="default_myplaces_style">
		<IconStyle>
			<Icon>
			</Icon>
		</IconStyle>
	</Style>
	<Placemark>
		<name>Eiffel Tower, France</name>
		<LookAt>
			<longitude>2.29426</longitude>
			<latitude>48.858052</latitude>
			<altitude>0</altitude>
			<range>265.415512</range>
			<tilt>0</tilt>
			<heading>22.250324</heading>
		</LookAt>
		<styleUrl>#default_myplaces_style</styleUrl>
		<Point>

<coordinates>2.29426,48.858052,0</coordinates>
		</Point>
	</Placemark>
</Document>
```

Since you'll be using Places quite a bit when you create Placemarks, Polygons, and Image Overlays, you should work with the folder structure in the Places tab enough to feel comfortable with it.

In the Places tab, collapse the Sightseeing folder.

Right-click the words "My Places" and choose Add, then Folder.

When the Google Earth—New Folder dialog comes up, name the folder "GIS" and click OK.

Now right-click the folder GIS and choose Add, then Placemark.

You'll see a new pushpin appear in the main viewer, wherever you happen to be located, and the Google Earth—New Placemark dialog comes up.

Name the placemark something imaginative, like "Test." Leave the Latitude-Longitude settings as they are.

In the Description box, type "First Placemark."

In the Style, Color tab, look at the settings for the label and the icon. Under "Icon," click the square box next to the word "Color" to bring up the color palette. Choose a color and click OK.

The Placemark icon in the view immediately changes to the new color.

Click the View and Altitude tabs, and look at the Latitude-Longitude, Tilt, Heading, and Altitude settings.

Click OK.

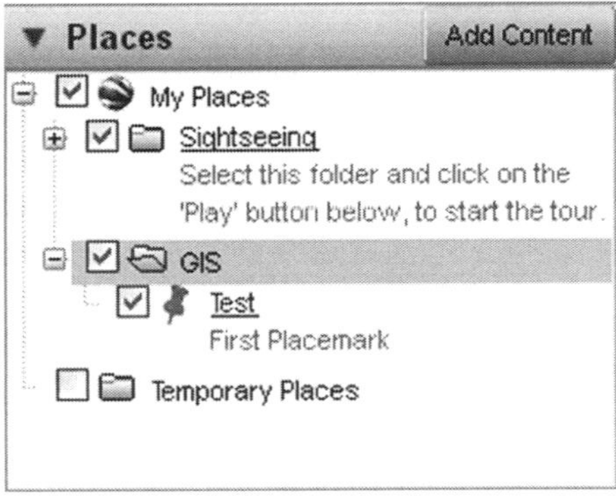

Now you have a new, permanent Placemark. Since you created it inside the My Places folder, it will be there next time you open Google Earth. If you had saved it in the Temporary Places folder, Google Earth would ask you upon closing if you wanted to make it permanent.

Navigating in Google Earth is pretty intuitive if you've used Google Maps, and it works *almost* like ArcGlobe and ArcMap. The left mouse rotates the globe in different directions, keeping the center of the earth as the center of rotation. If you hold down the middle scroll wheel and move the mouse in different directions, the earth rotates around a point on the surface at the center of the view. Holding down the center mouse wheel and pulling the mouse down or up is the quickest way to tilt the terrain. Holding down the right mouse button zooms and tilts at once; this is referred to as "swoop navigation" because when you fly in close enough, you travel along the terrain instead of zooming to the center of the earth. Scrolling the mouse wheel also zooms you in and out, but with finer control than holding down the right mouse button. Finally, you can set the globe spinning in any direction by holding down the left mouse button, swiping it across the main viewer, and letting up on the button before you stop moving the mouse. (It's the same movement you used in Chapter 3 when you set Animated Rotation in ArcScene.)

These effects can also be achieved with the Navigation Controls in the upper-right corner. The best way to get used to the controls is to practice with them, but there are a few settings worth knowing about that we'll look at in the meantime.

Step 5. Look at Some View Features

From the View Menu, choose Show Navigation. This setting lets you decide whether you want to see the Navigation Tools in the top-right corner all the time, none of the time, or just whenever you mouse over them.

Choose a setting that you like—I'm sticking with "Always" for now.

Again in the View menu, check on "Grid," "Status Bar," "Overview Map," and "Scale Legend."

The Scale Legend is what you'd expect, and very handy to have on. The Overview Map sits in the upper-left corner and helps keep you oriented as you pan and zoom in the main viewer. The location and extent of the main viewer is indicated by the size and position of the red square in the Overview Map.

The Status bar resides at the bottom of the main viewer, and shows you the coordinates, the terrain elevation, and your "eye altitude" at the cursor.

To see the Grid, zoom out far enough to get a worldly view. Depending on what geographical coordinate system Google Earth is using for display, you either see a latitude-longitude grid or a UTM grid. We'll look at how to change those settings in a few moments.

Step 6. Examine Some Application Settings

If you've spent any time using the mouse wheel to zoom in and out in Google Earth, you've noticed that, by default, scrolling the wheel away from you zooms you in, while scrolling the wheel toward you zooms you out. This is the opposite of the mouse action in ArcGIS products, and in Chapter 4 I showed you how to reverse the mouse wheel settings in ArcGlobe to match Google Earth's. But if, like me, you use ArcGIS 90 percent of the time, you'll be happy to know that you can also reverse the mouse wheel action in Google Earth. So without further ado:

From the Tools menu, choose Options.

Click the Navigation tab. Under Mouse Wheel settings, click Invert Mouse Wheel Zoom Direction.

Mouse Wheel Settings
Speed: Slow Fast
☑ Invert Mouse Wheel Zoom Direction

Click OK, and scroll in and out to try the new setting. (If you don't like it, of course, change it back.)

Open the Options dialog again from the Tools menu, and click the General tab.

Here you can decide if you want to see tooltips and start up tips, and what email application you'd like to use to send images or Placemarks, among other options.

Click the 3D View tab.

Important settings on the 3D View tab are the units that you want coordinates and elevation values to be displayed in, the vertical exaggeration (called "elevation exaggeration" here), and the quality of the terrain.

On the 3D View tab under the Show Lat/Long panel, change the units from Degrees, Minutes, Seconds to Universal Transverse Mercator.

Click OK, and notice the change in the Grid and the Status Bar in the main viewer.

From the Tools menu, choose Options again, and click the Cache tab.

The cache settings work the same way here as they do for ArcGlobe. You can change the amount of RAM and disk space that Google Earth uses to display imagery (when you're offline) and local KML data.

Close the Options dialog without changing the Cache settings.

Feel free to change back any of the settings in the Options dialog, and turn off or leave on the Grid, Overview Map, Navigation Controls, and so forth that might be cluttering your viewing experience. Continue to experiment with navigation and settings.

Exercise 5-2

Create a Polygon and Edit Its Properties Through Google Earth's Form Menus

Step 1. Start Google Earth and Fly to a Location

Double-click the Google Earth icon on your desktop, or start the program from Start > Programs > Google Earth.

On the Fly To tab on the Search panel, type "75001 Paris, France."

Press the Enter key.

You should arrive at the Louvre, in Paris.

Step 2. Turn 3D Buildings On and Off

This is a great time to see how users of Google Earth have created and shared their 3D building models. Many of the models are made in SketchUp, a free program that interfaces directly with Google Earth.

In the Layers panel, find the 3D Buildings layer, and make sure that it's turned on.

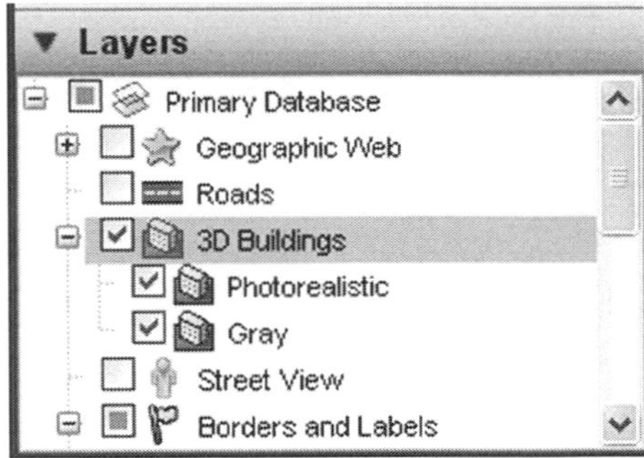

Zoom and navigate until you have a tilted view of the Louvre Museum and surrounding buildings. Turn the 3D Buildings on and off in the Layers panel, to compare the plain aerial and satellite imagery with user-created 3D buildings.

Step 3. Create a New Folder

If you did the last exercise, you have a GIS folder in the Places panel. If you don't have that, create one now in the following manner:

In the Places tab, collapse the Sightseeing folder.

Right-click the words "My Places," and choose Add, then Folder.

In the New Folder dialog, name the folder "GIS" and click OK.

(Make sure to save it in the My Places folder, not the Temporary folder. If you accidentally save it in the wrong folder, you can drag it to the correct folder, just as you would in ArcCatalog or Windows Explorer.)

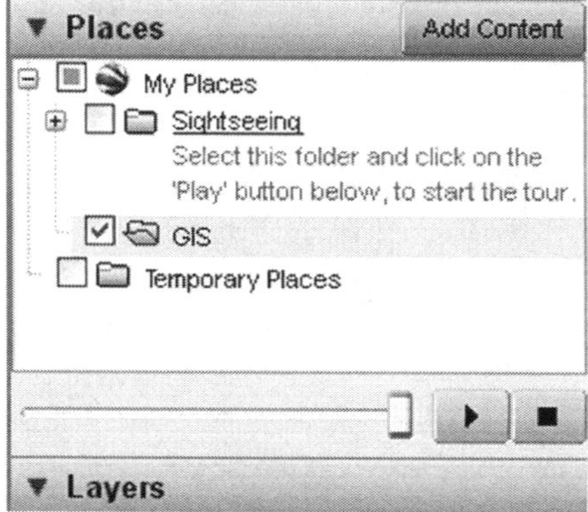

If you have a Placemark already in the GIS folder from the last exercise, you can delete it, or just ignore it.

Step 4. Create a New Empty Polygon

Zoom to the Tuileries and Carrousel Gardens west of the Louvre, and tilt the terrain so that it's orthographic—that is, so that you're looking straight down as if in a 2D view. You can do most of this by holding down the middle mouse button and dragging the cursor upward in the view. To make sure that you're completely in orthographic view, use the down arrow of the "Look joystick" on the Navigation Tools (the tool with the eye on it).

When you create new Placemarks, Polygons, Paths or anything else, Google Earth puts them in whichever folder is selected in the My Places panel.

Make sure that the GIS folder is selected in the My Places panel.

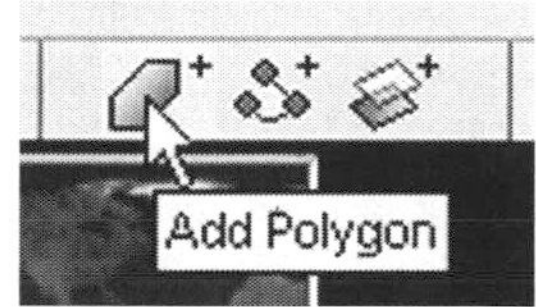

In the Toolbar at the top of the application, click the Add Polygon button.

In the New Polygon dialog, name the polygon "Gardens." For a description, type "Louvre gardens and grounds."

Select the Style and Color tab. Notice that the Line Width is 1.0, and the Opacity is 100% (completely opaque). The Area of the polygon is set to be filled and outlined, again with an opacity of 100%.

In the Area panel, change the color of the polygon fill by clicking the box next to "Color." When the color palette comes up, choose something bright that will stand out from a distance.

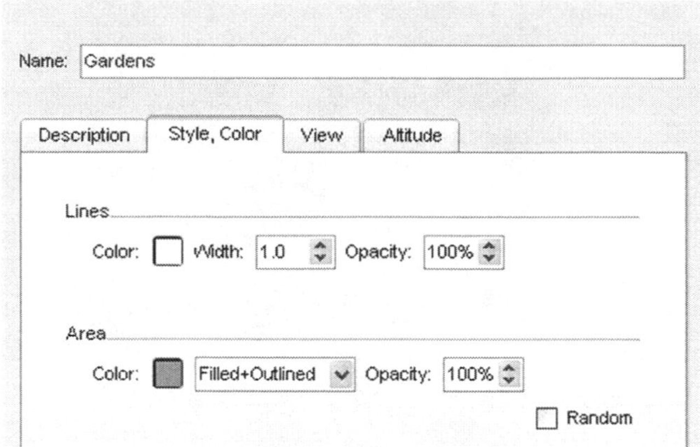

Select the View tab. Click the "Snapshot current View" button. This fills in the current values for the latitude, longitude, heading, tilt, and range (the distance from your camera eye to the earth).

Select the Altitude tab.

The first thing you see in the box next to Altitude is "0m." This is not a mantra, but in fact signifies "Zero Meters." As long as it's set to 0 meters, the polygon is clamped to the terrain. This is the default altitude setting.

Leave the settings in the Altitude tab alone for now. Google Earth gives you trouble if you alter the altitude settings before you draw the polygon.

Click OK to create the new empty polygon. Its name appears in the Places panel.

Step 5. Draw the Polygon

Right-click the new layer "Gardens" in the My Places\GIS folder, and choose "Properties."

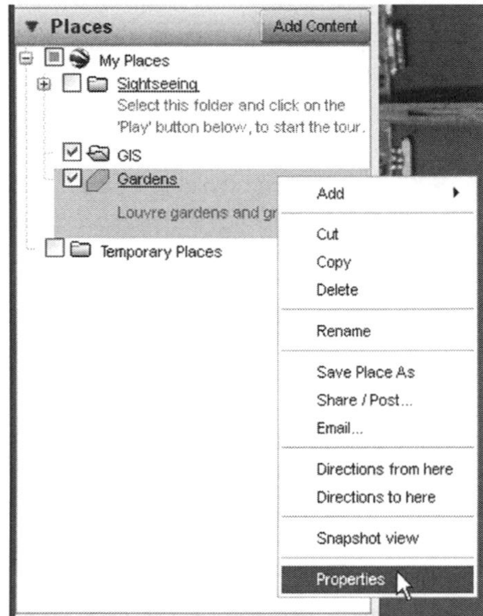

The Edit Polygon dialog comes up; it looks just like the New Polygon dialog. All editing of the polygon takes place while this dialog is open. (Also, it's very easy to *accidentally* edit the polygon while the dialog is open; for example, when your intention is to just innocently pan around in the view.)

Move the dialog to the side so you can see the main viewer.

Notice that while the Edit Polygon dialog is open, the cursor changes to a square drawing tool.

Click the left mouse button once at each of the four corners of the garden square. (If you hold down the left mouse button and drag, you'll get a free-form polygon, which is okay too but easy to make a mess of.)

If you make a mistake, click Cancel in the dialog and start over.

Once the polygon is made, you can select the four corners and drag a little bit to straighten the figure out.

When you're happy with the basic polygon shape, click OK in the Edit Polygon dialog.

Step 6. Add a URL to the HTML Popup

In the Places panel, click once on the word "Gardens."

The information that you typed in earlier—the name and description of the new feature—appears in a popup balloon on the polygon.

Right-click "Gardens," and select Properties.

On the Description tab, in the Description box, type the following URL:

www.louvre.fr/llv/musee/jardins_tuileries.jsp

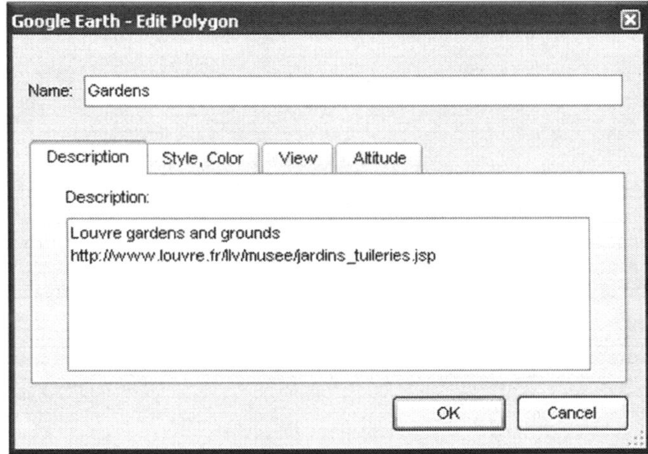

Click OK to close the Edit Polygon dialog.

Now the "Gardens" entry in the Places panel has a name, description, and a URL. Click the word "Gardens" again, once, to invoke the HTML popup.

In the popup, click the URL.

A new browser window appears at the bottom of the main viewer, displaying the website for the URL you specified.

When you've finished checking out the website, click the "X" in the upper-right corner of the new inset browser window to close it.

Step 7. Change the Polygon's Transparency and Height

I don't know about you, but what I have for my trouble so far is an ugly orange polygon obscuring the Tuileries Gardens in Google Earth. In order to make it a little more attractive, let's change some of the graphical effects.

Open the Edit Polygon dialog again by right-clicking "Gardens" in the Places panel and selecting Properties. Move it out of the way of the main viewer.

On the Style, Color tab, change the opacity of the Area to 40%, and change the Width of the Lines that make up the outline of the polygon to 5. If you like, choose a color besides white for the outline of the polygon.

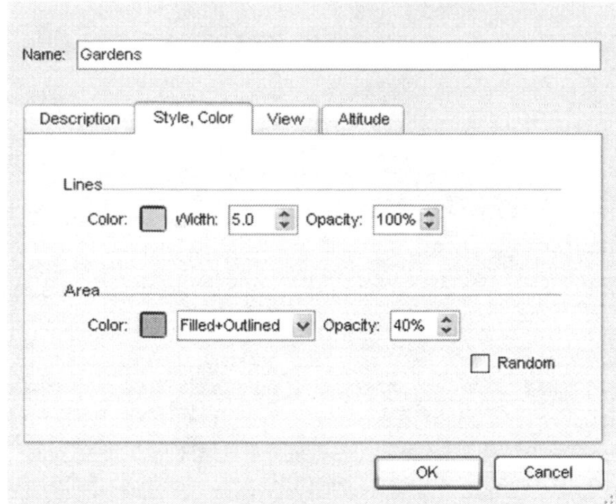

Notice that these changes take place in the main viewer immediately.

Click the Altitude tab.

The Altitude tab should perhaps be called the Elevation tab or Height tab, because it refers to settings that affect the height of the polygon, not the eye-altitude of the camera.

There are three altitude settings: Clamped to ground, Relative to ground, and Absolute.

"Clamped to ground" is the default elevation setting. The feature (be it Polygon, Path, Placemark, or Image Overlay) sits directly on the terrain. The value in the Altitude box for this will always be zero meters.

"Relative to ground" places the feature above the terrain, at whatever value in meters you set in the Altitude box.

"Absolute" should really be called "Relative to Sea Level" because that's what it means; it isn't absolute at all. You can change the value in the Altitude box and Google Earth will place the feature at that height in meters, but above sea level instead of above the terrain elevation.

The units in the Altitude box are always meters, no matter what Elevation display settings you choose in Google Earth's Tools/Options dialog (you saw the Options dialog in the last exercise).

Move the slider bar to the right, toward "Space." The Altitude reading reflects your changes in meters, and the dropdown menu changes from "Clamped to ground" to "Relative to ground."

Look at the three different settings in the dropdown menu, and slide the bar back and forth. When you're satisfied that you understand the concepts behind the settings, put in a value of 200 meters in the Altitude box, make sure the dropdown menu says "Relative to ground," and check the box to "Extend sides to ground." This will create a box instead of a floating flat polygon.

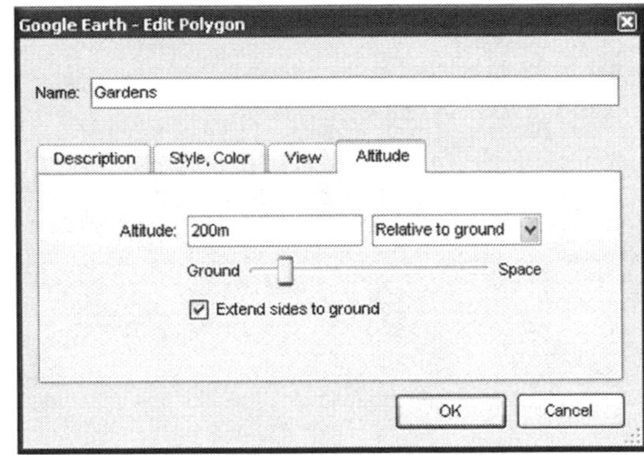

Click OK to close the dialog.

Take a turn around the new extruded polygon. Remember, holding down the center mouse button and dragging the cursor up and down tilts the terrain; holding the center mouse button down and moving the mouse from side to side revolves the view around the cursor's location.

In the next exercise, we'll take a peek at the KML code that makes up this polygon.

Exercise 5-3

Edit the Gardens Polygon Using KML

KML, much like HTML, is a markup language that tells Google Earth how to display information onscreen. It also has a similar structure to HTML in that all the elements participate in a "container" relationship. Each element is either nested within or contains other elements.

For example, a very simple container relationship can be seen in the following code:

```
<Placemark>
<address>75001 Paris, France</address>
</Placemark>
```

The <Placemark> element tells Google Earth to create a Placemark somewhere. That "somewhere" is defined in this case by the <address> element (as opposed to an element specifying latitude and longitude, for example). The <address> element is getting the information "75001 Paris, France" and so Google Earth puts the Placemark there.

Notice that the first part of an element has brackets around it, and the second part of the element has brackets with a forward slash around it, just like HTML. Elements always begin with an expression in a double-bracket set, and end with the same expression in a double-bracket-plus-slash set.

Here's an example of a slightly expanded container:

```
<Document>
        <name>test.kml</name>
        <Placemark>
                <name>Gardens</name>
                <description>Louvre gardens and grounds</description>
                <Point>
<coordinates>2.3369,48.86079999999999,0</coordinates>
                </Point>
        </Placemark>
</Document>
```

The Placemark element in the code above contains three other elements: name, description, and Point. (Capitalization is important in KML: "name" and "Name" are not the same.) The Point element contains one element: coordinates. The coordinates element, on the other hand, contains no additional elements; just the attribute information "2.3369, 48.86079999999999, 0."

What's great about KML is that it can be shared with other users who have Google Earth or Google Maps. All you have to do is save a feature—like our polygon feature—as a .kml file, email your friend the file, and they can place the feature in Google Earth on their local machine, just by double-clicking the .kml file. Or, if you make a particularly nice 3D model or Placemark with interesting photos and content attached, you can post the KML file to the 3D Warehouse or the Google Earth Community, so everyone can see it.

The code can be as simple or as complex as you want to make it. If you want to familiarize yourself more with the structure, a good starting place is Google Earth's tutorial and reference guide for KML, currently residing at:

http://earth.google.com/intl/en//userguide/v4/ug_kml.html

Step 1. Copy the Gardens KML Code to Notepad

If all went well in the last exercise, you should have a polygon that looks something like the one below, although with slightly different coordinates. (If you don't have your own polygon, I've put a copy of my polygon's code in your 3DDATA\Chapter05\Data folder for you to use. It's called gardens_bkup.kml.)

Open Notepad (not WordPad or Word). If you don't have a shortcut icon on your desktop for Notepad, I suggest you create one by clicking Start\Programs\Accessories, then right-clicking Notepad and choosing "Send To . . . . Desktop (create shortcut)."

Start Google Earth, and make the Places panel visible.

In the My Places\GIS folder, right-click the word "Gardens" and choose Copy.

In Notepad, click the Edit menu, then choose Paste.

From the File menu in Notepad, choose Save As.

Navigate to your 3DDATA\Chapter05\MyData folder.

From the Save as Type list, select All Files.

In the File name box, type "gardens.kml."

Be sure that you save the file as a .kml file, not as a .txt file. (If you wind up with a .txt file, fixing the problem is as easy as renaming the file extension.)

File name: gardens.kml Save
Save as type: All Files Cancel
Encoding: ANSI

Click Save.

Step 2. Examine the Container Relationship in the Polygon's KML Code

Google Earth puts in some of its own personal code that you might not use if you were creating the KML from scratch in Notepad. We won't go into most of it here, but we'll look at a few basic items before we make changes to the Gardens polygon from Notepad.

The first two items you see are lines stating that this code is in XML version 1.0 and KML version 2.2.

Container relationships have to have a "root" element, which in this case is the first <kml . . . > statement. Notice that at the very bottom of the file, you see the closing tag </kml>.

The "child" of the kml element is the Document element. At the bottom of the file you see the closing tag </Document>, just above </kml>.

The child of the Document element is the Folder element; you can see the closing tag </Folder> just above </Document>, at the bottom of the file.

Kml, Document, Folder, and Placemark are the four possible types of root elements. The kml element doesn't always have to be the outside container. For example, the extremely simple code you saw a couple of pages earlier:

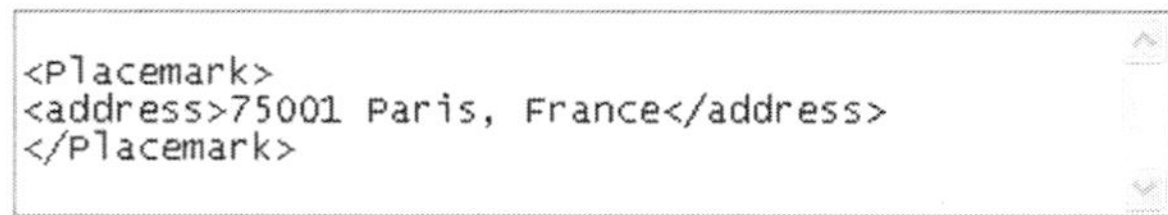

```
<Placemark>
<address>75001 Paris, France</address>
</Placemark>
```

can be pasted as is into Notepad, and then saved as a .kml file. Opening it in Google Earth will create a Placemark at the Louvre address, shown with the default yellow pushpin.

Step 3. Look at the <LookAt> Element

Scroll down until you see the expression:

```
<LookAt>
        <longitude>2.326870755323014</longitude>
        <latitude>48.863678</latitude>
        <altitude>0</altitude>
        <range>545.337734950213</range>
        <tilt>0</tilt>
        <heading>-64.45223560010049</heading>
        <altitudeMode>relativeToGround</altitudeMode>
</LookAt>
```

The LookAt element specifies two different things: the position of the feature being viewed, and the position of the "virtual camera" in relation to the feature. The camera position is the one that Google Earth flies to when you double-click Gardens in the Places panel.

The LookAt element houses six child elements. The latitude, longitude, and altitude elements specify the coordinates of the feature that the camera is looking at; in this case, the centroid of the polygon that we created. The altitude element is set to zero; zero means that the coordinate set we're looking at should be "Clamped to Ground," which is the default. (We'll get to the difference between <altitude> and <altitudeMode> in a moment.)

The range element specifies the distance in meters from the feature to the virtual camera. In my example, the camera is zoomed in pretty close—only 545 meters above the polygon.

The tilt element specifies the angle of the camera to the feature. In my case, the tilt is zero, and the camera is looking straight down. The tilt angle can be anything between zero and 90 degrees. Ninety would place the camera on the ground, 545 meters away from the feature.

The heading element specifies the azimuth, or compass direction, that the camera is facing, in degrees. Due north is 0 or 360 degrees; due west is 270 degrees. The heading in my case is –64 degrees. Now we come to <altitudeMode>.

In the last exercise, you changed the height value of the Gardens polygon by using the Altitude tab in the Properties dialog. You changed it from "Clamped to ground" (zero meters) to "Relative to ground" (200 meters, in our case). By doing that, you created the altitudeMode element. The altitudeMode element simply overrides the altitude element; both of them specify the height, or elevation, of the feature relative to the ground. Neither of them relates to the "altitude" of the camera; that's handled by the range and tilt elements working together.

Step 4. Look at the <Polygon> Element

Scroll down a couple more lines and focus on the expression:

```
        <Polygon>
                <extrude>1</extrude>
                <tessellate>1</tessellate>
                <altitudeMode>relativeToGround</altitudeMode>
                <outerBoundaryIs>
                        <LinearRing>
                                <coordinates>
2.323451935546174,48.86324758502238,200
2.329777298839389,48.86129792927846,200
2.331533159592027,48.86364392799987,200
2.325280375962748,48.86564157462371,200
2.323451935546174,48.86324758502238,200 </coordinates>
                        </LinearRing>
                </outerBoundaryIs>
        </Polygon>
```

The Polygon element specifies the coordinates that make up the boundaries of the feature. The "extrude" element takes a Boolean value that tells Google Earth whether or not the feature is connected to the ground when its altitude mode is other than "Clamped to ground." In this case, the value is 1, because we checked "Extend side to ground" on the Altitude tab in the Gardens polygon's Properties.

The "tessellate" element is another Boolean value that specifies whether or not the polygon follows the terrain.

Here again we see <altitudeMode>; in this case the value "Relative to ground" affects the height of the coordinate pairs for each vertex of the Polygon, not the height of the single coordinate pair of the LookAt element that we examined earlier.

The "outerBoundaryIs" element houses the "LinearRing" element, which in turn houses the coordinates for the vertices of the polygon. In my case, there are five pairs of coordinates for the four vertexes that I clicked—the first and last pair are identical, indicating that the beginning and end points of the LinearRing are the same.

Step 5. Change the Altitude of Some of the Polygon's Vertices

Many of the code elements we've looked at so far can easily be changed in the Edit Polygon dialog, in the Google Earth application. Some, however, can only be done through altering the KML code. For example, when you specified "200" as a value to extrude the polygon to, Google Earth set each vertex to a height of 200 meters. In the code, however, you can treat each vertex individually.

From the File menu in Notepad, click Save As.

Navigate to your 3DDATA\Chapter05\MyData folder.

Name the file "gardens_edit.kml" to keep it separate from the original KML file.

File name: gardens_edit.kml
Save as type: All Files
Encoding: ANSI
Save
Cancel

Click Save.

In the new gardens_edit.kml file, scroll down again until you get to the section of code specifying the LinearRing coordinates.

After each coordinate pair, you see the altitude value "200." Change the values of the middle three altitude values to something other than "200." You can follow my example below, or choose your own values.

```
                                <LinearRing>
                                        <coordinates>
2.323451935546174,48.86324758502238,200
2.329777298839389,48.86129792927846,500
2.331533159592027,48.86364392799987,700
2.325280375962748,48.86564157462371,50
2.323451935546174,48.86324758502238,200 </coordinates>
                                </LinearRing>
```

Click Save from the File menu, and close Notepad.

Step 6. Open Gardens_edit.kml in Google Earth

In Google Earth, turn off "Gardens" in the GIS folder.

From the File menu in Google Earth, select Open.

Navigate to your 3DDATA\Chapter05\MyData folder, and select gardens_edit.kml.

The new polygon opens in the Temporary Places folder. The changes you made to the altitude values of the polygon's vertices should be evident.

Experiment some more with the new polygon and its KML code. When you're finished, save the new polygon in the GIS folder, and close Google Earth.

CHAPTER **6**

Raster Surface Models

Raster Interpolation

In Chapter 1, you learned about the cell structure of raster data. You also learned that in a thematic raster, samples of the phenomenon are used to estimate the unknown cell values in the model. This estimation process is called interpolation, and it is used to create models of measurable phenomena such as elevation, rainfall, air pollution, snow depth, noise, and contamination.

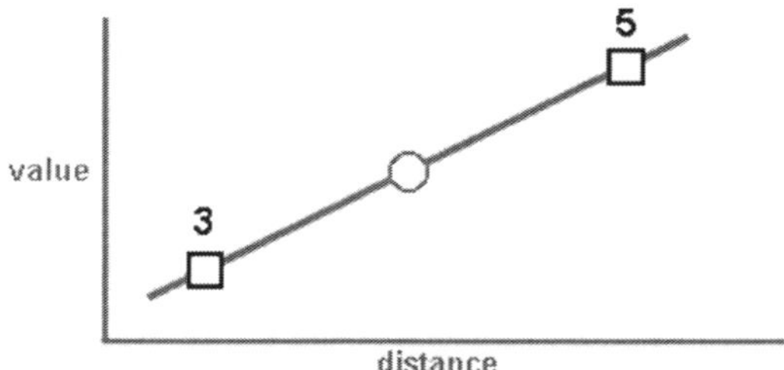

An example of linear interpolation. The center dot lies midway between two known values, 3 and 5, so its value is estimated to be 4

As you might imagine, interpolating values across a surface is more complicated than estimating them along a straight line. In the grid below, an elevation surface has been interpolated from a set of points. Each point represents a location where the elevation has been measured. The resulting raster is a prediction of what the elevation is at any location on the actual surface. The more points accurately sampled, the better the prediction.

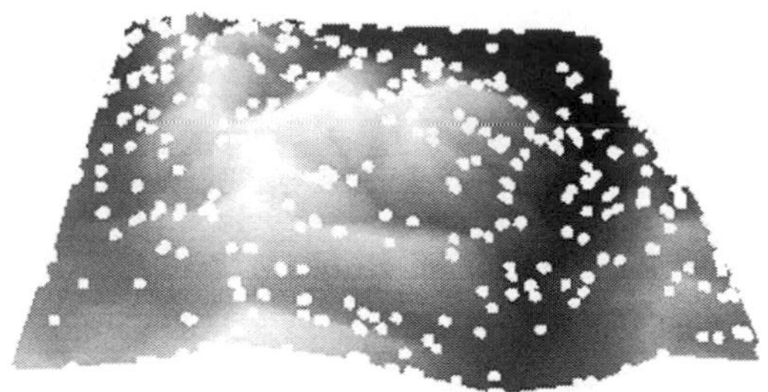

Points on surface show where the samples were taken. Extra samples were taken where the slope changes rapidly, in order to create a more accurate model

Spatial interpolation can be used to create a surface from just a few sample points. However, more sample points are better if you want a detailed model. Input points can be randomly, deliberately, or regularly sampled. For an area where the surface changes sharply or the phenomenon is concentrated, a cluster of sample points may be necessary.

The central notion behind spatial interpolation is that points near each other are more alike than those further apart. Therefore, the value at any location should be estimated using nearby points. Most interpolation methods apply this principle by giving the nearest sample points the most influence when estimating an unknown cell value. 3D Analyst offers several interpolation methods for creating raster surfaces from point data, including Inverse Distance Weighted (IDW), Spline, Natural Neighbors, Kriging, and Trend.

		25		
			9	
	5	?		
		3		33
19				

Numbered cells indicate known values. In estimating the value of the unknown center cell, an interpolation method using spatial autocorrelation will use all six sample points but will give more influence to the nearest ones. Here, the darker cells indicate the most influential sample points

Sample Size

Most interpolation methods let you control the number of sample points used to estimate cell values. If you limit your sample to seven points, the interpolator will use the seven nearest known values to estimate each unknown cell value. You can also control the sample size by defining a search radius. A fixed search radius will only use the samples contained within it. If not enough sample points are found within the search radius, you can use a variable search radius that expands until the specified sample size is found.

Interpolation Barriers

What if your surface contains a sharp break in terrain, such as a river bank or a cliff? If you interpolate values on either side, the sharp break will be smoothed over. One interpolation method (IDW) lets you include barriers in your analysis. The barrier works like a force field, essentially interpolating two separate surfaces. Unknown values on one side do not take any samples from the other side for analysis. Interpolation barriers can add considerably to the time it takes to compute unknown values. One way to speed up the process is to use as few vertices in the barrier feature as possible.

If terrain contains a sudden change in elevation, you can set an interpolation barrier to keep samples at one side of the break from affecting values on the other side of the break

IDW, Spline, Kriging, and Natural Neighbors interpolation methods work differently, and which method will produce the most accurate surface will depend on the phenomenon you're modeling and the distribution of sample points. No matter which method you use, the more evenly distributed input points you have, the more accurate the results.

Inverse Distance Weighted (IDW) Interpolation

IDW is the least complicated interpolation method. Unknown cell values are calculated by averaging the values of sample points in the vicinity of each cell. The closer a sample point is to the center of the cell being estimated, the more influence it has in averaging the cell's value. A specified number of points, or all of the points within a given radius, can be used to compute the value of each output cell.

IDW works best for closely packed, consistently spaced sample point sets. It tends to flatten out surface variation rather than preserve it. It's good for measuring phenomena such as sound, whose distribution falls off sharply with distance. It's not so good for measuring phenomena like air pollution, because it can't take into account any prevailing trends in the data, such as wind. Because IDW averages values, the surface model won't pass through the sample points, and no estimates will be made above the maximum or below the minimum sample values.

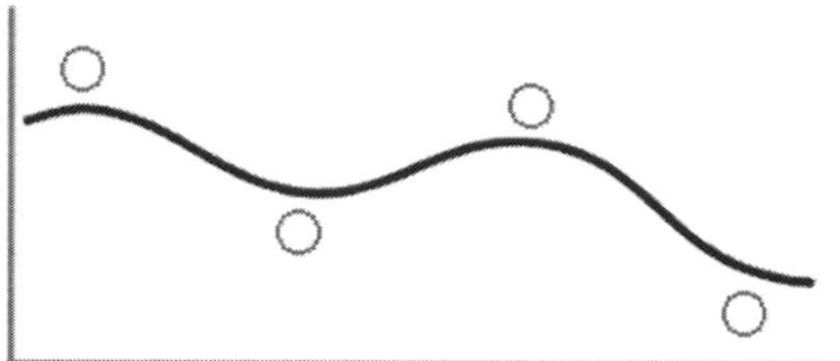

Because IDW averages values, a surface created with it won't pass through or exceed the value range of the sample points

That being said, IDW offers a number of adjustable parameters that other interpolation methods don't. IDW comes with a Power setting that adjusts the relative influence of the sample points. A Power setting of 0 gives all samples (within the specified search radius) almost equal influence in estimating the output point. As the Power setting increases, the influence of sample points falls off more rapidly with distance. The output cell values are more localized, so the new surface has more detail. A power of 2 is most commonly used.

You can also choose the type of radius used to control the number of sample points. A Fixed Radius setting uses a constant distance, including all of the sample points found within that radius to interpolate an unknown cell value. A Variable Radius setting lets you specify the number of points that will be averaged. The radius for each interpolated cell will be different, because it has to expand until it finds the specified number of sample points. It's better to use a variable radius when the density of sample points varies significantly from one area to another. If you have areas with sparsely distributed input points, you can put a limit on the radius, so that if the number of points isn't reached inside the maximum distance of the radius, the interpolator will use fewer points in the calculation.

IDW also lets you use polylines to limit the search for sample points. The lines may represent a cliff, a river, or some other interruption in a landscape. As discussed earlier, only sample points on the same side of the barrier as the unknown cell will be used to estimate it.

Spline Interpolation

The Spline interpolator, instead of averaging values between sample points, creates a surface that passes exactly through them. This is useful if you want to be able to estimate values that are below the minimum or above the maximum values found in the sample data.

Spline is best for surfaces that vary gradually. If samples are close together and have extreme differences in value, it can overshoot estimated values because it uses slope calculations (change over distance) to figure out the shape of the surface. Terrain or phenomena that changes suddenly, such as a cliff face or a fault line, are not well represented by a smooth curving surface. In those cases, you might be better off using IDW interpolation.

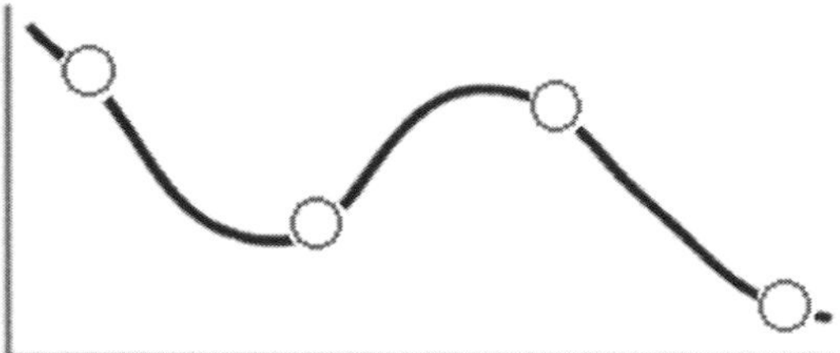

The Spline interpolator creates a surface that passes through each sample point. The surface values may surpass the sample values

There are two Spline methods, Regularized and Tension. Regularized Spline creates a smooth, gradually changing surface with values that may lie outside the sample data range. Tension spline forces the estimates to stay closer to the sample data, resulting in a tighter, less elastic surface.

Kriging Interpolation

Kriging is similar to IDW in that it weights nearby point samples to estimate an unknown value. In IDW, however, this weight depends solely on the distance to the unknown location. Kriging assumes that the distance and direction between sample points indicate a spatial relationship in the surface, so the weights are based not only on the distance between the measured points and the unknown location but also on the overall arrangement among the measured points. It adapts its calculations to the data by analyzing all of the data points to find out how much correlation they exhibit, and then factors it in the weighted average estimation. Kriging involves several steps before it creates the surface, including exploratory statistical analysis of the data and variogram modeling. It's a very complicated interpolation engine, most appropriate when you know there is a distance or directional bias in your data. It is often used in soil science and geology. A great deal more information can be found in the ArcGIS 3D Analyst Help, and in the 3D Analyst user manual.

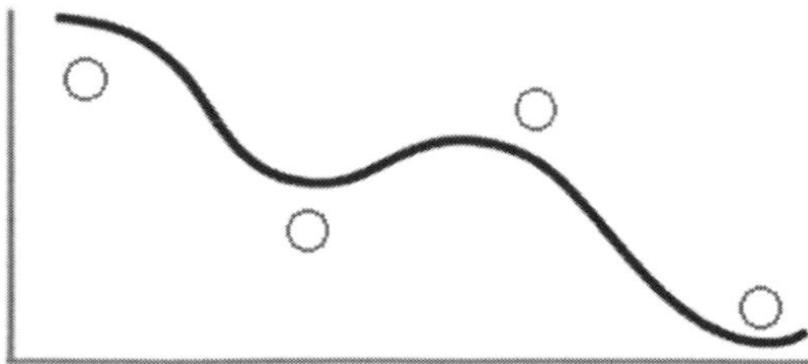

A surface created with Kriging can exceed the value range of the sample points, but won't pass through them

Two oft-used Kriging methods are ordinary and universal Kriging. Ordinary Kriging assumes that there is no particular trend in the data, while Universal Kriging assumes that there is an overall movement across the surface. For example, you may know that wind is coming from one direction, or that the terrain slopes consistently. Universal Kriging should only be used when you know there is a prevailing trend in your data.

Natural Neighbors Interpolation

Like IDW, Natural Neighbors interpolation is a weighted-average method. However, instead of estimating a point's value using sample cells weighted by their distance, the Natural Neighbors interpolator first performs a Delaunay triangulation. A Delaunay triangulation is what is used to create a TIN from sample points. We'll cover that in more detail later when we talk about TINs, but suffice to say that a Delaunay triangulation connects all the sample points so that each point is a node in a triangle, and all triangles are contiguous and nonoverlapping.

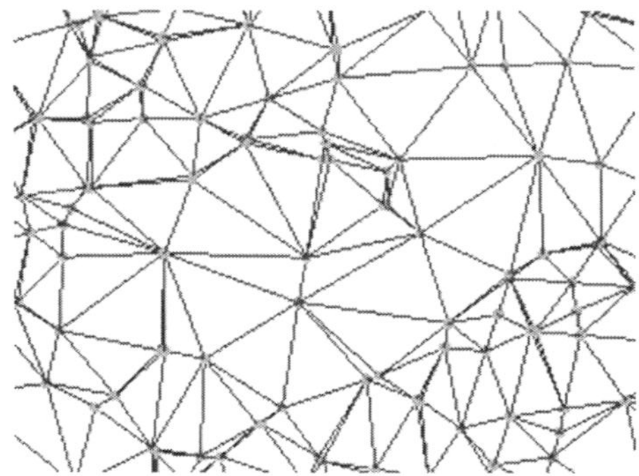

Delaunay triangulation connects irregularly located points with lines to form a network of contiguous, nonoverlapping triangles

When the triangles are formed, Natural Neighbors creates a set of Voronoi polygons around the sample points. Voronoi polygons, also called Thiessen polygons, are formed by drawing the perpendicular bisector of each of the triangle lines so that each polygon bounds the region that is closer to one point than to any other point.

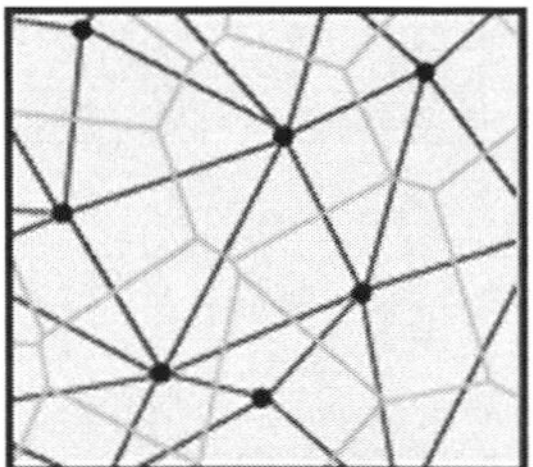
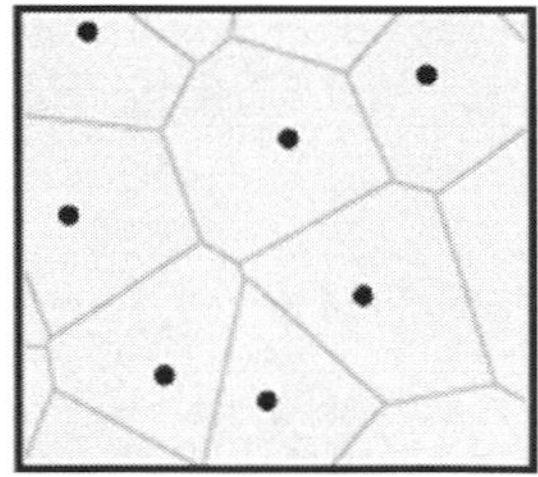

On the left is a set of sample points connected by Delaunay triangulation, with the Voronoi polygons in the background. On the right, the Voronoi polygons drawn by bisecting each triangle side. The result is that each polygon encloses the area that is closer to a point than to any other point

Finally, the raster surface is interpolated using the sample data points that are "natural neighbors" of the cell centers. The value of an estimated location is a weighted average of the values of the natural neighbors. Since the output is a raster, the estimation locations are a regularly spaced array equal to the number of raster cells.

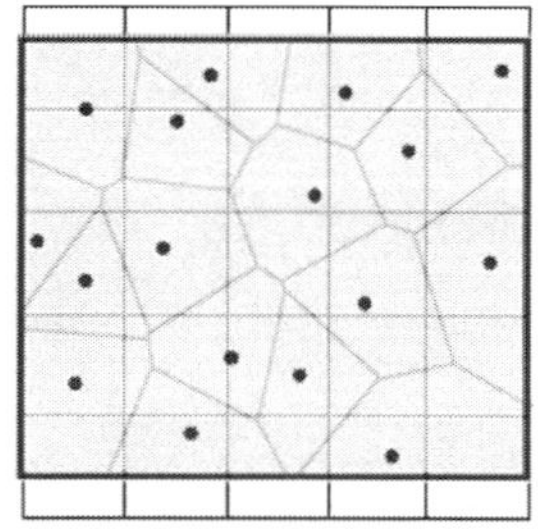

For each cell in the output raster, a weighted average of the cell center's natural neighbors is calculated

Natural Neighbors is most suitable for sample data points that are unevenly distributed. Because it uses a weighted average, it doesn't require specific parameters such as radius, number of sample points, or weights.

Trend Interpolation

Trend, as its name implies, creates a smooth surface that varies gradually. It applies a polynomial to the sample points, and makes a surface that passes through some of them but for the most part tries to interpolate distances between sample points such that those points that are higher than the surface and those that are lower will, when added up, not be that different in their respective sums. Trend allows you to set a polynomial order between 1 and 12. An order of 1 fits a flat plane to the points; higher orders will add bends to the surface and increase the overall accuracy.

Remember, when you use any of these interpolation methods, you should have a clear understanding of your data and of the phenomenon being modeled.

Raster Reclassification Methods: Slope, Aspect, Hillshade, and Viewshed

Now that you know a bit about how rasters are interpolated from sample points of elevation, let's look at some common ways that cell values can be reclassified.

When you reclassify a raster, you replace an existing set of cell values with a new and usually smaller set of values. 3D Analyst provides engines for calculating commonly desired raster surfaces from elevation values that reveal the slope, aspect, hillshading, and viewsheds of an area.

For a raster of phenomena other than elevation, you may want to assign new values in order combine it with other rasters for further analysis, to change classification schemes, or to assign standards of preference, priority, or sensitivity.

Reclassification should not be confused with resampling, which transforms the cell size or coordinate space of a raster. Reclassifying changes cell values but not the resolution of the raster (the number of cells per square unit of area in the grid).

Slope

Slope is the steepness of a surface, measured as the change in surface value over distance. Slope can be measured in degrees from horizontal (0 to 90) or as percent slope, which is the rise divided by the run, times 100. A slope of 45 degrees equals 100 percent slope. Measures of slope in degrees approach 90 degrees (vertical), and measures of slope in percent approach infinity.

In a raster, slope is calculated as the maximum rate of change in value between a cell and its eight closest neighbors. Areas of steep slope are prone to erosion and landslides.

Aspect

Aspect is the compass direction that a slope faces. Values are measured clockwise in degrees from 0 (due north) to 360 (again due north). A slope with an aspect value of 270, for example, faces west. If you walked down that slope, you'd be walking west. (If you walked up the slope, you'd be walking east, so aspect describes the downslope direction that a cell faces.) Flat areas get an aspect value of -1.

In a raster, aspect is figured for each cell by estimating the fit of a plane to the z values of that cell plus its eight neighbors. The aspect of the plane becomes the aspect value of the cell in a new raster.

Aspect modeling is used in many applications, such as determining the amount of solar heating a building will receive, how much sun vegetation will get, and how fast snow will melt.

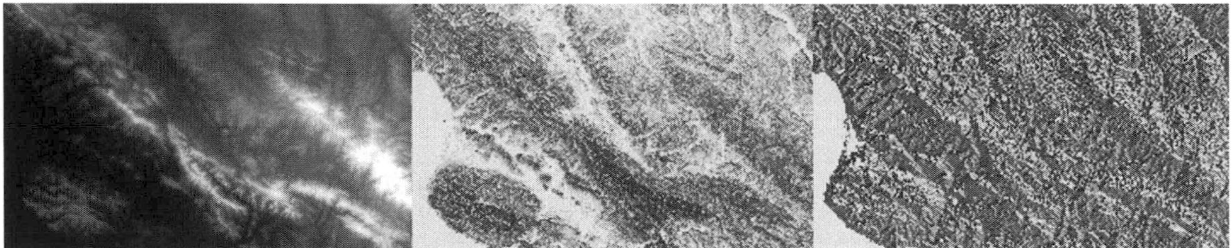

The center and right panels are slope and aspect rasters calculated from the elevation raster on the left. Aspect is usually symbolized with a bright, primary color scheme: one color for each direction, plus gray for flat areas

Hillshade

Hillshade is the pattern of light and dark that a surface shows when lit from a particular angle. It's used to model the amount of sun and shadow that an area receives at different times of day, and to increase the perception of terrain relief in a two-dimensional surface.

The Hillshade function in 3D Analyst is like slope and aspect, in that it calculates a new raster based on the sun angle and steepness of each cell on the surface. The value of each hillshaded cell ranges from 0 (no light) to 255 (brightest light).

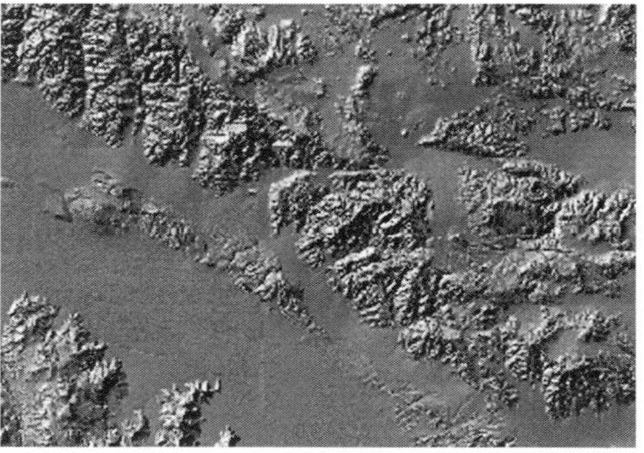

A hillshade raster calculated with the sun's azimuth at 315 degrees (shining from the northwest) and an altitude of 45 degrees above the horizon

Viewshed

Viewshed identifies the areas of a surface that can be seen from one or more observation points. It is useful in aesthetic applications, such as determining the value of real estate, and practical applications such as the placement of telecommunications towers.

The observation points can be features in a point layer, or the vertices in a line layer. Each cell in the viewshed raster stores the number of observers who can see it. If you have only one observation point, each cell that can be seen from that point gets a value of 1. All cells that can't be seen from that point get a value of 0.

If more than one observer point is used, each visible cell in the raster stores the number of points that can see it. Viewshed analysis assumes that the observer's field of view has no horizontal or vertical constraints, and no distance limits.

A viewshed raster derived from a shapefile (the white observation point) and an elevation raster

More Reclassification

Slope, aspect, hillshade, and viewshed are the most common rasters that are derived from height values. With 3D Analyst, you can use an elevation raster or a TIN as the source for those heights.

Rasters—whether they contain elevation values or not—can be reclassified for a myriad of analysis applications. You may want to replace cell values with updated information, give certain cells a NoData value, or classify values together before converting the raster to a vector format. Often, several rasters are reclassified to a common scale so that they can be combined to model a larger spatial problem.

For example, a raster of soil types might be reclassified to create a raster of soil pH values, values in a slope raster might be reclassified into values of suitability, or a raster of wildlife habitat might be reclassified into values of sensitivity or preference. For more information on raster reclassification and its uses, see *About Analyzing Surfaces* and *Reclassifying Your Data* in the ArcGIS Desktop Help.

Long ago in Chapter 1, you learned that a difference is often struck between image rasters and thematic rasters. Thematic rasters, in turn, are often divided into two categories: those that model categorical data and those that model continuous data. Although the differences between categorical and continuous phenomena can sometimes be argued, continuous data generally represents a phenomenon that varies continually over a surface, such as elevation, precipitation, or pollution concentration. Categorical data—also called discrete data—represents a phenomenon with known boundaries that can be placed into classes, such as a raster of landuse categories.

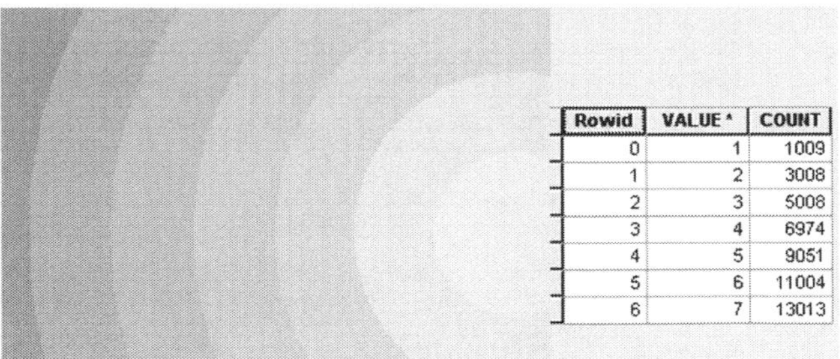

Cell values in this integer raster are classed into categories of buffer distance. The buffers could also be represented by vector polygons

When you reclassify continuous data, you replace ranges of values with individual values. When you reclassify categorical data, cell values are replaced on a one-to-one basis. For example, in a soil habitat analysis, you might want to replace soil types, represented by numeric codes, with soil pH values. You can also reclassify discrete data into ranges. For example, if you were investigating places to build a new park, you could reclassify a raster that contains 10 landuse categories to create a raster containing only three preference categories of low, medium, and high.

Now that you've been introduced to raster interpolation and classification methods, it's time to create some raster surface models, calculate slope, hillshade, and aspect, and create viewsheds from observer points.

Exercise 6-1

Interpolate a Terrain Surface with Spline

In this exercise, you'll use the Spline interpolator to create elevation models of San Luis Obispo, California.

Step 1. Start ArcScene and Add Data

Open ArcScene and click the Add Data button.

Navigate to your 3DDATA\Chapter06\Data folder.

Add slocitypoints.shp to the scene.

Step 2. Examine the Shapefile

Zoom in to the layer and have a look. Slocitypoints is a point shapefile of elevations samples in a portion of San Luis Obispo County.

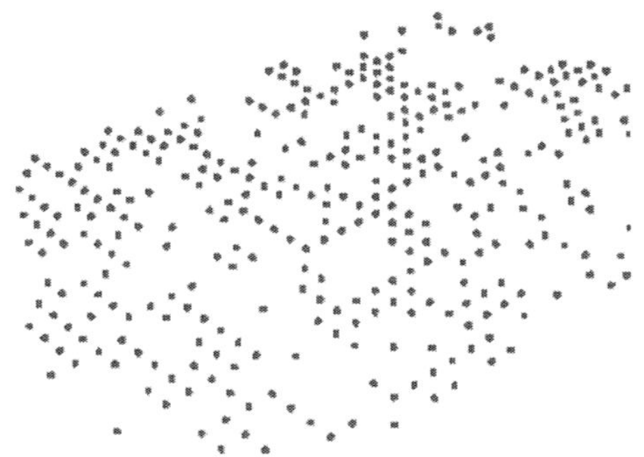

Step 3. Change the Scene to Orthographic View

From the View menu choose View Settings, and click Orthographic (2D) View. This will make it easier to compare differences in the raster surfaces that you are about to create.

Close the View Settings dialog.

Step 4. Open the Point Layer's Attribute Table

In the Table of Contents, right-click slocitypoints and choose Open Attribute Table.

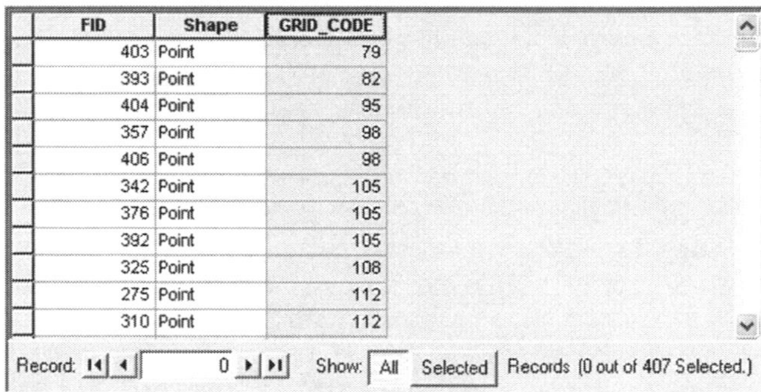

FID	Shape	GRID_CODE
403	Point	79
393	Point	82
404	Point	95
357	Point	98
406	Point	98
342	Point	105
376	Point	105
392	Point	105
325	Point	108
275	Point	112
310	Point	112

Right-click the GRID_CODE field and choose Sort Ascending.

The attribute table contains 407 records. Scroll through it and look at the sample elevation values in the GRID_CODE field. They range from about 80 to 1200 feet. This dataset is a subset of a larger point shapefile; the actual sample values were taken over a larger area and ranged further in value, and these samples were selected at random from the larger set. Some of the highest and lowest values across the area, therefore, are not represented.

Close the attribute table.

Step 5. Set the Analysis Environment

If the 3D Analyst menu isn't visible, click the View menu, then Toolbars, and check 3D Analyst.

From the 3D Analyst menu, choose Options.

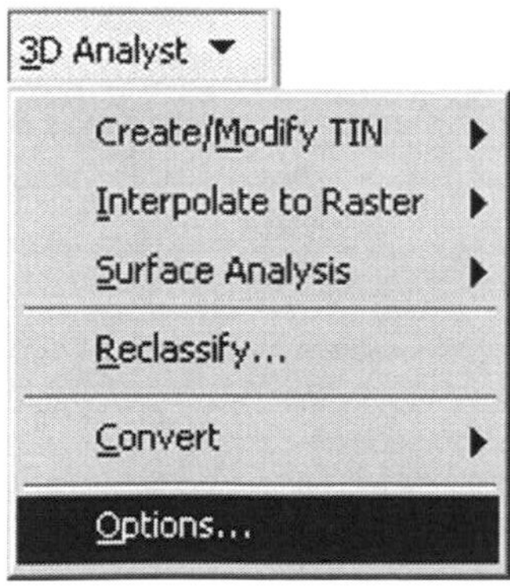

Click the General tab.

Use the Browse button to set the Working directory to 3DDATA\Chapter06\MyData, and set the Analysis mask to None.

The Working directory is where analysis results are stored. Most rasters created in 3D Analyst are temporary—that is, they are not saved to disk. This includes rasters created through interpolation, reclassification, and surface analysis. Exceptions to this rule are the Natural Neighbors interpolator and the TIN-to-Raster conversion function. Rasters created in that fashion are permanently saved to a directory that you specify.

If you want to make a raster permanent, you can change its temporary status when you create it, saving it either in the Working directory or in some other location. You can also right-click on a temporary raster in ArcScene's Table of Contents and choose Make Permanent or Data/Make Permanent, depending on which version of the software you have. (You may have to customize the Data context menu to include the Make Permanent command.) Finally, if you save a map or scene document, any temporary rasters it contains should be saved permanently to your Working directory. Rasters saved in this way are given a name followed by a number. The name describes the type of raster, and the number identifies the raster individually; for example, "IDW8," "SPLINE5," or "RCLASS3."

Still in the Options dialog, click the Extent tab. Set the analysis extent to Same as Layer "slocitypoints."

On the Cell Size tab, choose "As Specified Below" from the Analysis cell size menu, and type 98 into the cell size box.

Click OK.

Step 6. Interpolate a Surface with Spline

You'll use the Spline interpolator because it can estimate values above and below the range of sample points. There are no abrupt elevation changes, either, so Spline will work well for this surface.

From the 3D Analyst menu, choose Interpolate to raster, and click Spline.

Set the Z value field to GRID_CODE.

The Output cell size should be 98, and the Spline type should be Regularized.

Leave the other parameters at their default settings. The higher the Weight setting in Regularized Spline, the smoother and more rubbery the surface. Values for this parameter have to be equal to or greater than zero. Typical values are 0, .001, .01, .1, and .5. The default value is .1.

Click OK.

Spline

Input points: slocitypoints

Z value field: GRID_CODE

Spline type: Regularized

Weight: 0.1

Number of points: 12

Output cell size: 98

Output raster: <Temporary>

OK Cancel

The interpolator builds an elevation model based on the values in the GRID_CODE field.

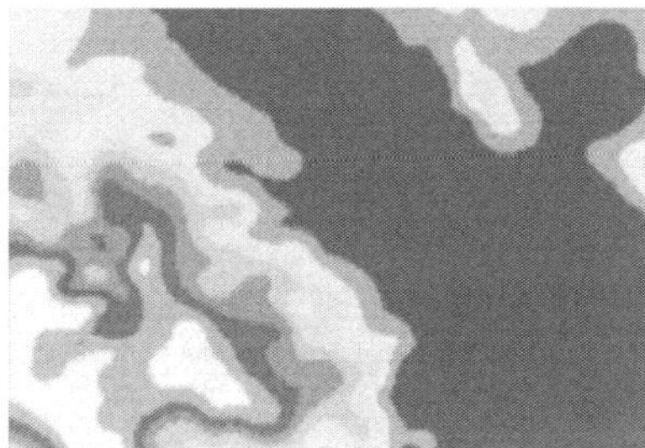

Step 7. Examine the New Raster

In the Table of Contents, notice that the elevation values range from about 55 to 1284 feet. This range is greater that the span of values you saw in the attribute table of slocitypoints.

Click the name "Spline of slocitypoints" to select it, then click it again to make a highlighted box around the name. Rename it "Regular Spline 0.1."

Turn off Regular Spline 0.1 in the Table of Contents, and collapse its legend.

Step 8. Change the Weight and Run Spline Again

You're going to increase the weight setting, amplifying the bend of the surface between the sample points and creating a looser, more flexible interpolation.

From the 3D Analyst menu, choose Interpolate to raster, and click Spline.

Make sure the Z value field is set to GRID_CODE.

Change the Weight setting from 0.1 to 1, and click OK.

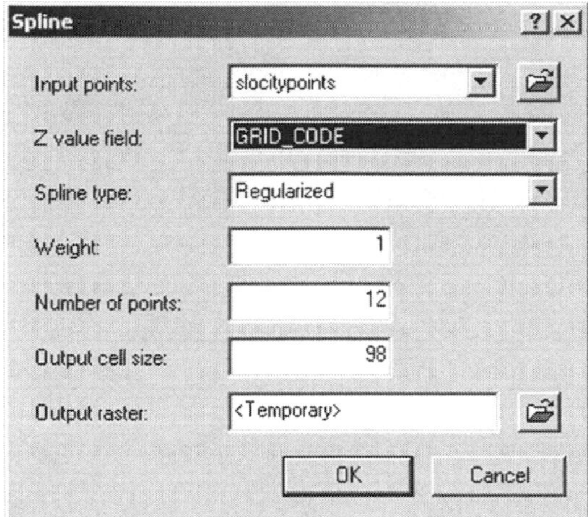

A second layer called Spline of slocitypoints is added to the scene.

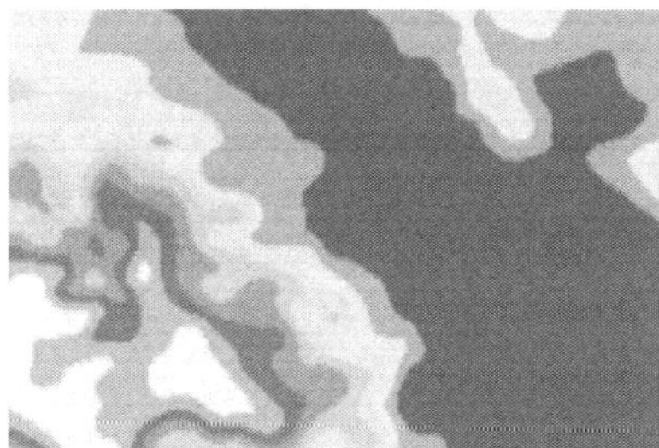

Compare the two surfaces by turning the layers on and off. The differences are slight, but you can see changes, particularly in the lower elevations symbolized in dark green. (The layers draw in the reverse order of their order in the Table of Contents, although if you were to save the ArcScene document, close it, and reopen it, they would draw in the conventional order.)

Rename the new surface Regular Spline 1 and collapse its legend. Turn off all of the layers.

Step 9. Run Tension Spline

Tension spline creates a less elastic surface than Regularized spline, keeping the minimum and maximum estimations closer to the sample data points. The tension curve is flatter than the regularized curve, meaning that there is less overall variation in the final values.

From the 3D Analyst menu, choose Interpolate to raster, and click Spline.

Make sure the Z value field is set to GRID_CODE.

Change the Spline type to Tension, but leave the rest of the parameters at their defaults.

Click OK.

A third layer called Spline of slocitypoints is added to the scene.

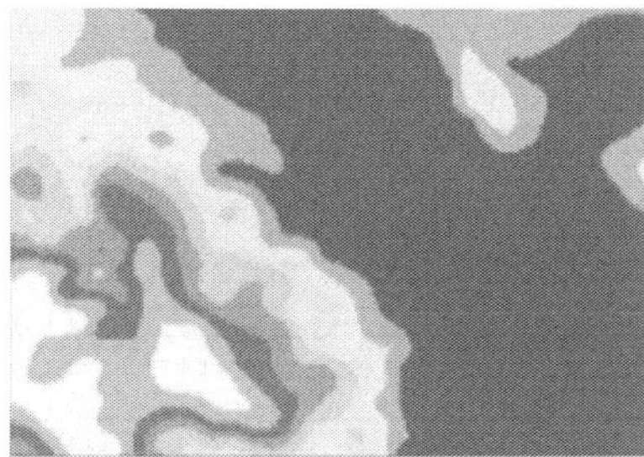

Turn the layers on and off to compare the three surfaces. The values in the Tension spline correspond more closely to the original range of sample values in the slocitypoints shapefile. The Regular Spline 0.1 surface falls in between the tighter Tension spline and the more elastic Regular Spline 1.

Rename the new surface Tension Spline 0.1, and collapse its legend.

Step 10. Set Layer Properties

From the View menu click View Settings. Change the projection from Orthographic to Perspective.

Close the dialog.

In the Table of Contents, turn on all the layers.

Right-click Tension Spline 0.1, and choose Properties.

Click the Source tab, and look at the Data Source panel. The temporary rasters that you've been creating have the name "SPLINE" followed by a number, and are saved in the Working Directory that you specified when you set up the analysis environment in Step 5. If you made more than three rasters during this exercise your raster may have a higher number, but in any case the most recent raster you created will have the highest number.

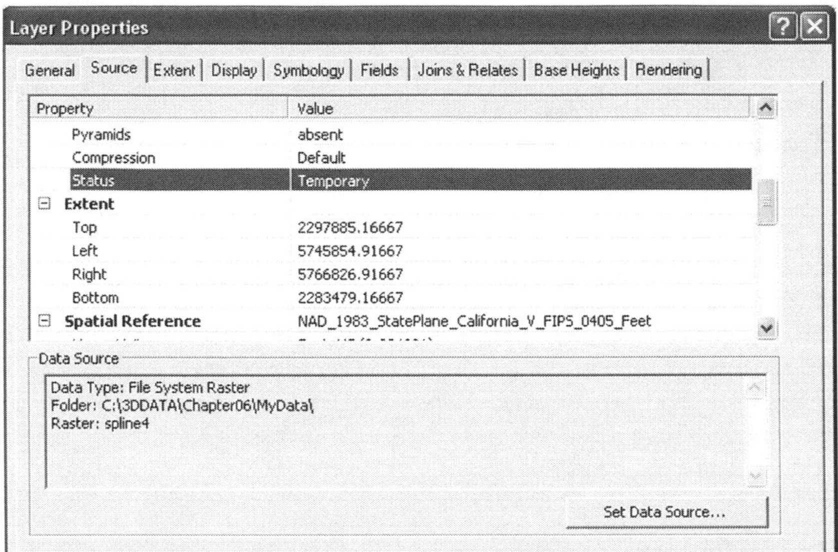

Now click the Base Heights tab. Click "Obtain heights for layer from surface," and make sure that the appropriate raster is selected.

In the Offset panel, type 3000.

On the Rendering tab, click to shade areal features, and click OK to close the Layer Properties dialog.

Open the Layer Properties for the other two layers. Set their base heights to the appropriate rasters and shade their areal features. Give Regular Spline 1 an offset of 1500. Leave the offset for Regular Spline 0.1 set to 0.

Step 11: Set Scene Properties

Now you have a nice sandwich of surface rasters, and you can tell which one you're looking at when you turn them on and off. But let's exaggerate the vertical to make the surfaces a little more interesting.

In the Table of Contents, double-click Scene Layers to bring up Scene Properties.

Click the General tab, set the Vertical Exaggeration to 2, and click OK.

Zoom to the full extent of the scene, and turn on all three spline surfaces.

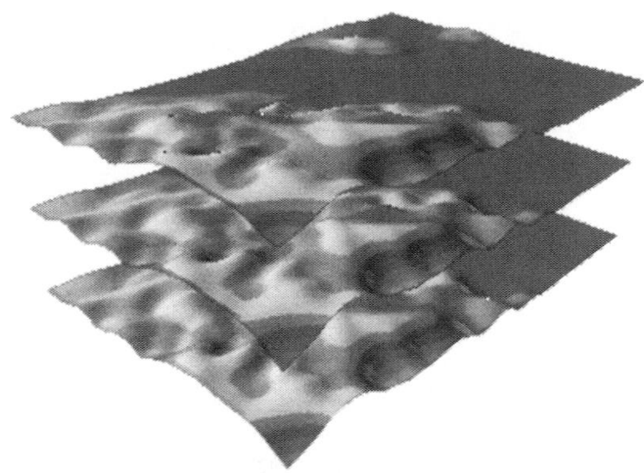

Step 12: Navigate the Scene and Save the Document

Zoom in closely to the three surfaces, turning layers on and off. When you're finished, save the ArcScene document under a name of your choice in your 3DDATA\Chapter06\MyData folder. Then close ArcScene.

Exercise 6-2

Interpolate Terrain with Inverse Distance Weighted and Natural Neighbors

Step 1. Start ArcScene and Add Data

Open ArcScene, and click the Add Data button.

Navigate to your 3DDATA\Chapter06\Data folder.

Add slocitypoints.shp to the scene.

Step 2. Set the Analysis Environment

From the 3D Analyst menu, choose Options.

On the General tab, set the Working directory to 3DDATA\Chapter06\MyData.

Set the Analysis mask to None.

On the Extent tab, set the analysis extent to Same as Layer "slocitypoints."

On the Cell Size tab, choose "As Specified Below" and set the cell size to 98.

Click OK.

Step 3. Change the Scene to Orthographic View

From the View menu choose View Settings, and click Orthographic (2D) View.

Close the View Settings dialog.

Step 4. Interpolate a Surface with IDW

From the 3D Analyst menu, choose Interpolate to Raster, and click Inverse Distance Weighted.

Set the Z value field to GRID_CODE.

The Output cell size should be 98, and the Power setting should be 2.

Leave the other parameters at their default settings, and click OK.

Inverse Distance Weighted

Input points: slocitypoints

Z value field: GRID_CODE

Power: 2

Search radius type: Variable

Search Radius Settings

Number of points: 12

Maximum distance:

Use barrier polylines:

Output cell size: 98

Output raster: <Temporary>

OK Cancel

A new elevation model called IDW of slocitypoints is added to the scene.

Step 5. Set Base Heights for Slocitypoints

In the Table of Contents, turn on slocitypoints, and double-click it to bring up its Layer Properties.

On the Base Heights tab, click "Obtain heights for layer from surface," and choose the new IDW raster from your Chapter06\MyData folder.

Click OK.

In the last exercise, you studied three surfaces of this area created with the Spline interpolator. There are a couple of things to notice about the IDW surface. One, the range of interpolated cell values is within the range of values from the original point shapefile, because IDW averages values. Two, the IDW interpolator has left rings around the sample points, like little grease spots.

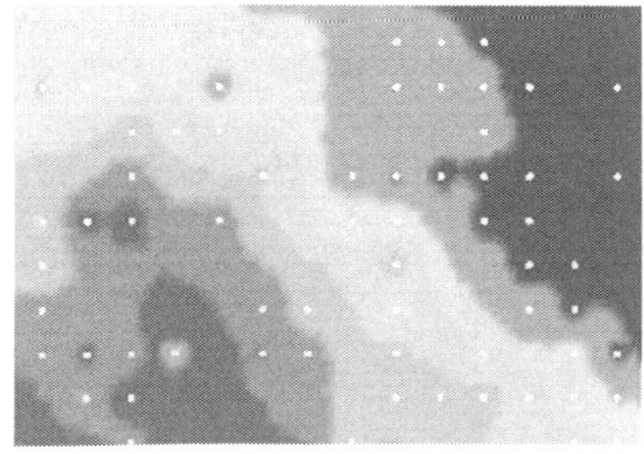

The sample points are exerting too much local influence because the Power setting is too high. A Power setting of 0 gives near and far points almost equal influence in estimating a cell's output value. As the Power setting increases, the influence of nearby sample points is given greater weight. The output cell values become more localized and less averaged, and the resulting surface will have more detail. In this case, the detail is exaggerated.

In the Table of Contents, click on IDW of slocitypoints once to select it, then again to make a highlighted box around the name. Rename it IDW Power 2.

Turn off IDW Power 2 in the Table of Contents, and collapse its legend.

Step 6. Change the Power Setting and Run IDW Again

From the 3D Analyst menu, choose Interpolate to Raster, and click Inverse Distance Weighted.

Make sure that the Z value field is set to GRID_CODE.

This time, change the Power setting to 0.4.

The Output cell size should be 98, and other parameters should be at their default settings.

Click OK.

A new elevation surface is added to the scene. It looks quite different from the IDW surface with a Power setting of 2. The local influence of the sample points has been decreased, resulting in a much more generalized surface.

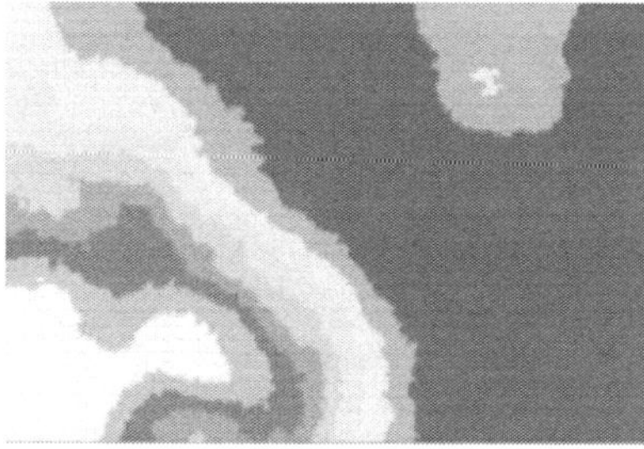

Rename the surface IDW Power 0.4, and collapse its legend.

Turn off all of the layers in the Table of Contents.

Step 7. Interpolate the Surface with Natural Neighbors

Since the Natural Neighbors interpolator is designed for irregularly sampled values, it should work well for slocitypoints—some areas of which are sampled densely and some areas of which have no samples at all.

From the 3D Analyst menu, choose Interpolate to Raster, and click Natural Neighbors.

The Natural Neighbors interpolator does not have many parameters; you can't specify a radius, a weight, or a number of sample points to use. And unlike IDW and Spline, it is not affected by most of the Analysis environment parameters that you set in the Options dialog. For example, you can't control the extent of the output raster, and you can't use a mask.

Make sure that the Height Source field is set to GRID_CODE. Save the output raster in your 3DDATA\Chapter06\MyData folder.

Let the Natural Neighbors dialog choose the rest of the parameters.

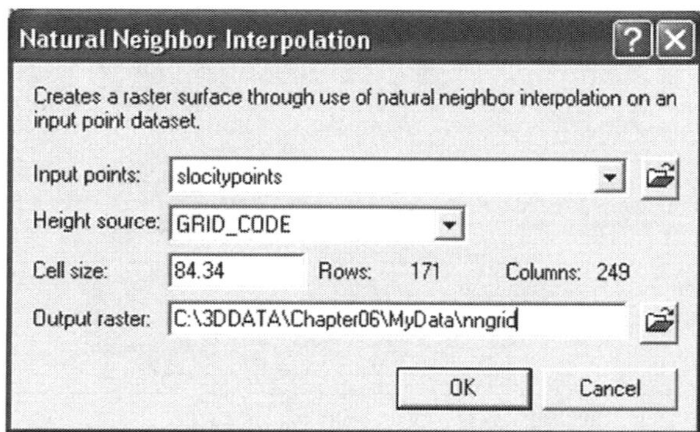

Click OK.

Wow! The surface looks pretty weird. Unlike the rasters you interpolated earlier, the Natural Neighbors surface doesn't come with default elevation color scheme. The symbology is also stretched instead of classified.

Step 8. Symbolize the Natural Neighbors Surface

In the Table of Contents, double-click nngrid to bring up its Layer Properties.

Click the Symbology tab. In the Show box, click Classified.

In the Classification panel, change the number of classes to 10.

Click the Classify button and change the classification method to Equal Interval. Click OK to close the Classification dialog.

Right-click the color ramp, and uncheck Graphic View. Scroll down and select the Surface color ramp.

On the Base Heights tab, click "Obtain heights for layer from surface," and choose the NNGRID raster from your Chapter06\MyData folder.

On the Rendering tab, check the box to shade areal features.

Click OK to close the Layer Properties dialog.

Step 9. Change View Settings and Scene Properties

From the View menu, choose View Settings. Change the projection back to Perspective.

Close the dialog.

In the Table of Contents, double-click Scene Layers to bring up the Scene Properties.

On the General tab, change the Vertical Exaggeration to 2.

Click OK.

Now the Natural Neighbors raster looks more recognizable, its estimations falling somewhere between the overlocalization of IDW Power 2 and the overgeneralization of IDW Power 0.4. Zoom in to take a look at the new surface.

Step 10. Close ArcScene and Save the Document

Feel free to experiment with the various interpolators. When you're finished, save the ArcScene document under a name of your choice in your 3DDATA\Chapter06\MyData folder.

Exercise 6-3

Calculate Hillshade and Aspect

In this exercise, you will create hillshade and aspect rasters from a DEM of a portion of San Luis Obispo County.

Step 1. Start ArcScene and Add Data

Open ArcScene and click the Add Data button.

Navigate to your 3DDATA\Chapter06\Data folder.

Add slocity_dem to the scene.

Slocity_dem is a Digital Elevation Model covering a larger area of San Luis Obispo County than you worked with in the last exercise.

Step 2. Look at slocity_dem's attributes

In the Table of Contents, right-click slocity_dem and choose Open Attribute Table.

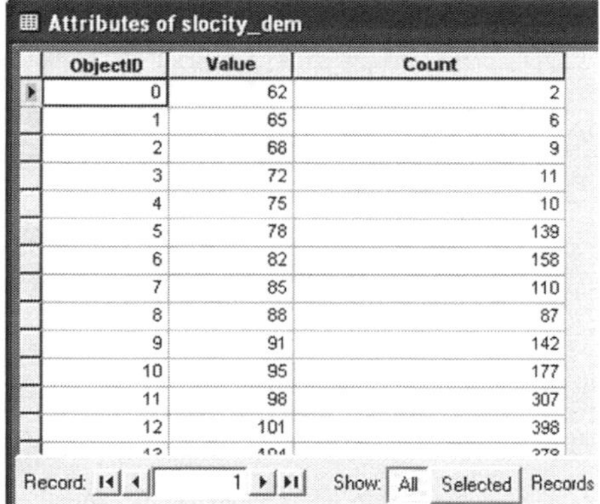
Attributes of slocity_dem

ObjectID	Value	Count
0	62	2
1	65	6
2	68	9
3	72	11
4	75	10
5	78	139
6	82	158
7	85	110
8	88	87
9	91	142
10	95	177
11	98	307
12	101	398

Slocity_dem has 805 records, with elevation values ranging from 62 to 2710. It was converted to an integer raster from a DEM with floating point values.

Close the attribute table.

Step 3. Set the Analysis Environment

From the 3D Analyst menu, choose Options.

On the General tab, set the Working Directory to 3DDATA\Chapter06\MyData.

Set the Analysis Mask to None.

On the Extent tab, set the Analysis extent to Same as Layer "slocity_dem."

On the Cell Size tab, set the Analysis cell size to Same as Layer "slocity_dem."

Click OK.

Step 4: Calculate Hillshade

On the 3D Analyst menu, choose Surface Analysis, and then Hillshade.

The input surface should be set to slocity_dem.

You may remember that azimuth is the compass direction of the sun, and that altitude is the angle of the sun above the horizon. Here, the default azimuth value of 315 degrees places the sun in the northwest, and an altitude of 45 puts it halfway between the horizon and the zenith.

As discussed earlier, a hillshade calculates the illumination of each cell on a surface from a hypothetical light source. Cells can be given values from 0 (black) to 255 (white). The Model Shadows option intensifies shadowing by assigning a value of 0 to all cells that are in the shadow of another cell, instead of giving them some value between 1 and 255. Shadow modeling is used to identify cells that will be in the shadow of another cell at a particular time of day.

Click the Browse button next to Output Raster and navigate to your 3DDATA\Chapter06\MyData folder. Call the hillshade "hillshd1" and click Save.

Leave the other parameters at their default values. The Hillshade dialog should look like this:

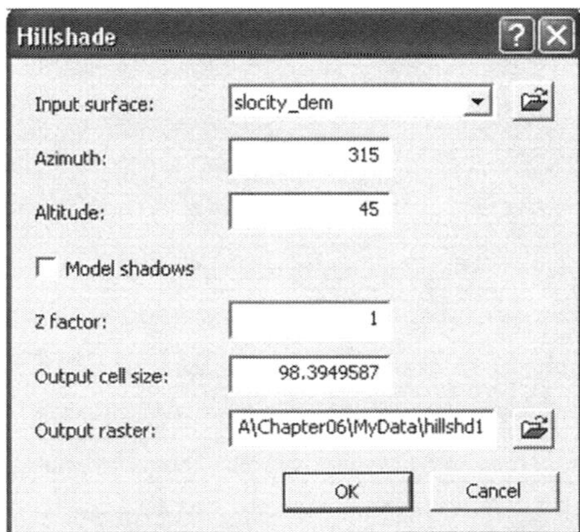

Click OK.

The new hillshade raster is added to the scene.

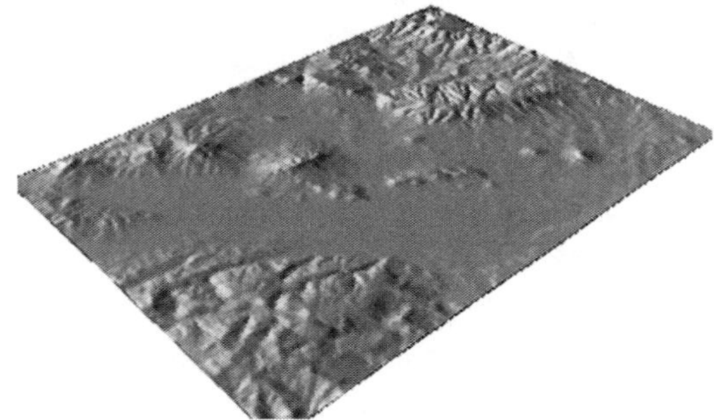

Notice that even though no base heights have been set for either the DEM or the hillshade raster, it looks three-dimensional. Hillshading is often used to add the effect of terrain relief to 2D data. By placing a semi-transparent elevation raster on top of a hillshade, you can create an even more realistic image of the landscape.

Step 5. Set Symbology and Transparency for Slocity_dem

In the Table of Contents, double-click slocity_dem to open its Layer Properties.

On the Symbology tab, right-click the Color Ramp and uncheck Graphic View. Scroll down and select the ramp called Yellow to Green to Dark Blue.

On the Display tab, type 40 in the Transparency box.

Click OK to close slocity_dem's Layer Properties.

Step 6. Set Drawing Priority for Hillshd1

In the Table of Contents, double-click hillshd1 to bring up its Layer Properties.

Click the Rendering tab. In the Effects panel, choose a drawing priority of 2.

Click OK.

The partially transparent DEM draws over the hillshade, giving an impression of topographic relief.

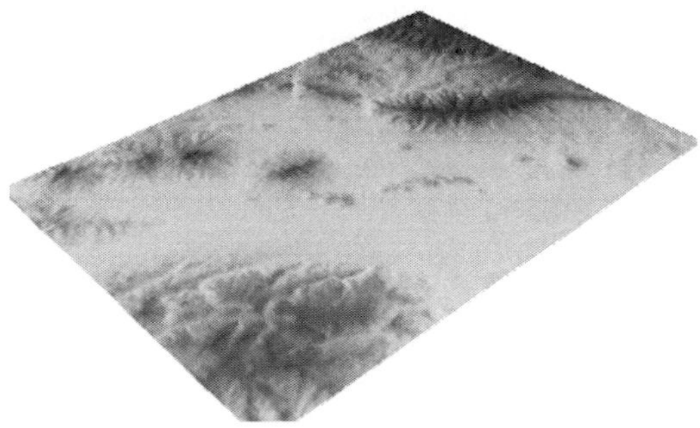

Turn hillshd1 off and back on in the Table of Contents to see the effect of the combined rasters. This way of showing relief is commonly used in orthographic displays—for example, in ArcMap where 3D viewing isn't available.

Step 7. Calculate Aspect

As you may remember, aspect identifies the steepest downslope direction from each cell to its neighbors. It can be thought of as slope direction or the compass direction a hill faces.

From the 3D Analyst menu, choose Surface Analysis, and then Aspect.

Make sure that the input surface is set to slocity_dem. (An aspect raster calculated from hillshd1 would look pretty odd, although you can feel free to try it out at the end of the exercise.)

Name the output raster aspect1, and save it in your 3DDATA\Chapter06\MyData folder.

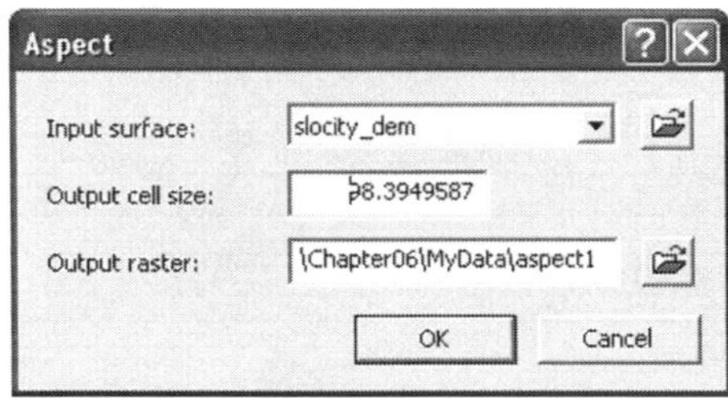

Click OK.

The new aspect raster is added to the scene.

The aspect raster is automatically symbolized with nine colors for the eight compass directions plus gray for flat areas. Red symbolizes North, yellow is East, medium blue is Southwest, and so on.

Step 8. Set Base Heights for All Three Layers

In the Table of Contents, double-click aspect1 to bring up its Layer Properties. (If ArcMap asks whether you want to compute unique values, click Yes.)

On the Base Heights tab, click "Obtain heights for layer from surface" and select slocity_dem.

Click OK.

Do the same for slocity_dem and hillshd1, setting the base heights in each case to slocity_dem.

Step 9. Set Vertical Exaggeration

In the Table of Contents, double-click Scene Layers.

Set the Vertical Exaggeration to 2, and click OK.

Zoom in and navigate around the surfaces, turning them on and off to see the different effects that hillshade, hillshade plus elevation symbology, and aspect create. Also try changing the transparency levels for aspect1 and slocity_dem.

Step 10. Save the Document

When you're finished experimenting, give the scene a name of your choice and save it in your 3DDATA\Chapter06\MyData folder. Then close ArcScene.

Exercise 6-4

Calculate Slope

Knowing the slope of an area is important for many applications, from modeling stream runoff to predicting mudslides to delineating flood plains. In this exercise, you'll create a slope raster of terrain around Santa Margarita Lake in California. You'll also reclassify a raster, and then use it as an analysis mask to keep the lake from being included in the slope model.

Step 1. Start ArcScene and Add Data

Open ArcScene, and click the Add Data button.

Navigate to your 3DDATA\Chapter06\Data folder.

Add the sanmar raster to the scene.

Sanmar is Digital Elevation Model of the area surrounding Santa Margarita Lake in San Luis Obispo County.

Step 2. Set the Analysis Environment

From the 3D Analyst menu, choose Options.

On the General tab, set the Working Directory to 3DDATA\Chapter06\MyData.

The Analysis mask should be set to None.

On the Extent tab, the Analysis extent should default to "Intersection of Inputs." On the Cell Size tab, the Cell size should be set to "Maximum of Inputs."

Click OK.

Step 3. Calculate Slope

From the 3D Analyst menu, select Surface Analysis, then Slope.

Click the Browse button next to the Output raster pathname. Navigate to your 3DDATA\Chapter06\MyData folder, and call the new raster "sm_slope."

Leave the other parameters as they are. You could calculate a raster that showed either percent or degree of slope, but in this case we'll stick with the latter.

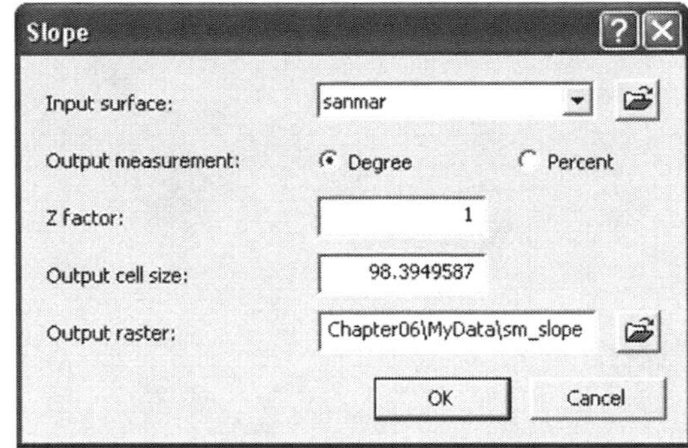

Click OK.

A new slope raster is added to the scene.

In the Table of Contents, turn off sanmar.

Zoom in and take a look at the slope raster. Each cell in the terrain is symbolized by its degree of slope, with steeper slopes in dark red and flat regions in dark green.

When you're done, collapse the legend for sm_slope.

Step 4. Add a Mask Layer

Zoom to the full extent of the scene.

Click the Add Data button, navigate to your 3DDATA\Chapter06\Data folder, and add san_mask to the scene.

Turn off sm_slope in the Table of Contents.

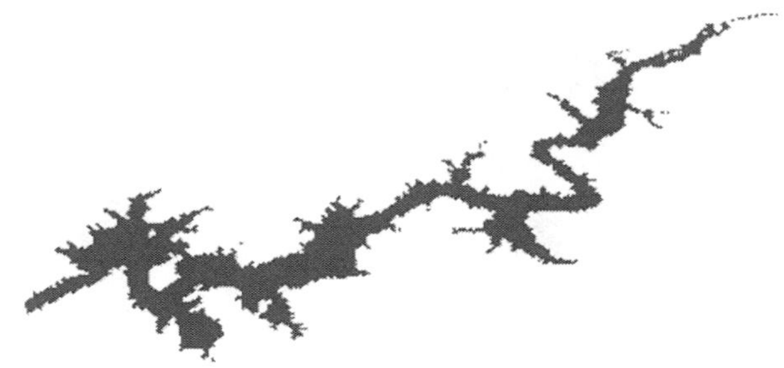

Step 5. Look at San_mask's Attributes

Right-click san_mask in the Table of Contents, and choose Open Attribute Table.

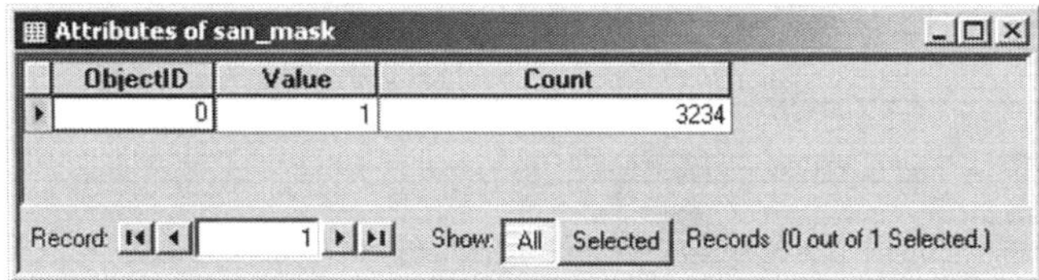

San_mask is a raster covering the same area as sanmar, but with only two cell values: 1 and NoData. The NoData values are not shown in the attribute table; all that you can see is that there are 3,234 cells with a value of 1. These cells correspond to Santa Margarita Lake.

Close the attribute table.

Step 6. Symbolize San_mask

3D Analyst uses a transparent "color" to symbolize NoData values. You can give them a color, however, to get a better idea of the geographical extent of san_mask.

Double-click san_mask to bring up its Layer Properties.

Click the Symbology tab.

Click the dropdown arrow next to "Display NoData as," and choose a bright green such as Chrysophase.

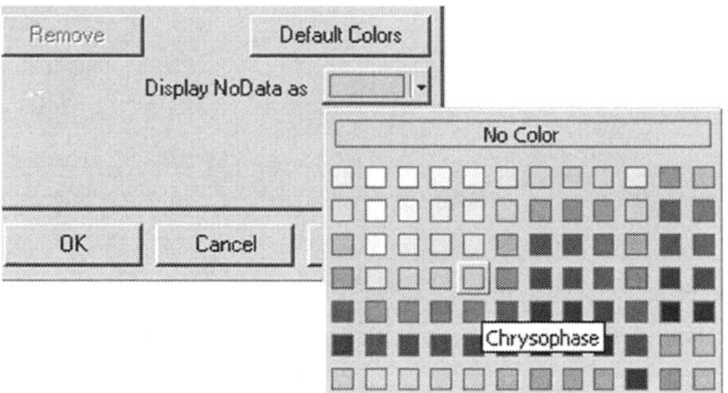

Click OK to close the Layer Properties dialog.

Now you can see san_mask's NoData values in the scene.

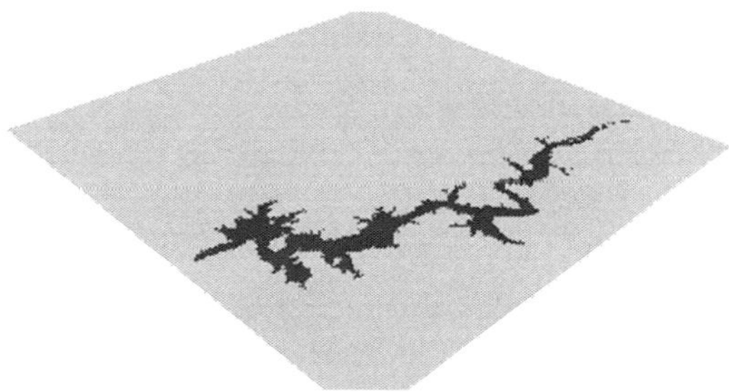

An analysis mask is a layer used to keep certain areas from being processed during the interpolation or reclassification of a new raster. It can be either a vector layer (point, line, or polygon) or a raster. Wherever the raster is covered by the mask, it is processed. Areas outside the mask, or areas inside the mask valued "NoData" are not processed.

What you'd like to do is create a new slope raster, but you don't want slope values calculated into the lakes.

In our case, san_mask has values of 1 wherever there is water, and values of NoData for the rest of the surface. If you used the mask as it is, the new slope raster would only be processed where the lake is, which is not what you want. So you're going to create a new mask layer, reclassified so that the land has a value of 1 and the water body has a value of NoData. This way your new slope raster will be calculated for the land only, leaving a blank area of NoData to demarcate Santa Margarita Lake.

Step 7. Check the Analysis Environment

When you're reclassifying or interpolating a surface, it's always good to check the analysis environment. 3D Analyst holds onto some of the previously set parameters, such as cell size, even if they were last set in an entirely different scene. If you create a surface and it comes out with the wrong extent or cell size, or is saved in the wrong location, it probably has to do with the settings in the analysis environment.

From the 3D Analyst menu, choose Options.

On the General tab, make sure that the Working Directory is set to 3DDATA\Chapter06\MyData.

The Analysis mask should be set to None. (You're making the mask in this step, not using it yet.)

On the Extent tab, set the Analysis extent to Same as Layer "san_mask."

On the Cell Size tab, also set the Analysis cell size to Same as Layer "san_mask."

Click OK to close the Options dialog.

Step 8. Reclassify San_mask to Create a New Mask Layer

From the 3D Analyst menu, select Reclassify.

Set the Input raster to san_mask.

In the New Values panel, type "NoData" to replace the old value of 1, and type "1" to replace the old value of NoData.

Call the Output raster no_lake, and save it in your 3DDATA\Chapter06\MyData folder.

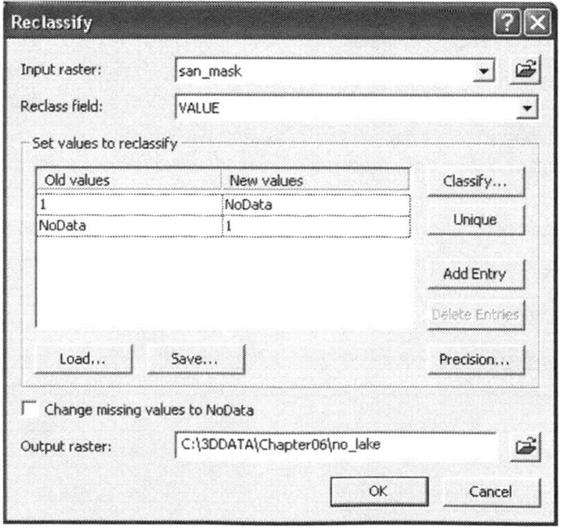

Click OK.

The new raster is added to the scene.

In the Table of Contents, turn off san_mask so that you can see no_lake undisguised.

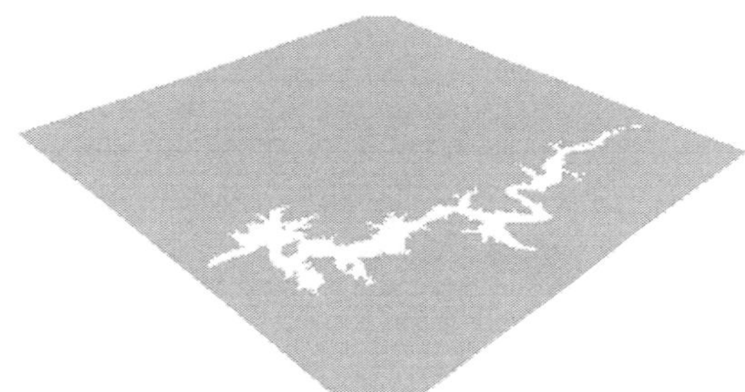

No_lake has the value 1 wherever there is land, and the value NoData wherever there is water. By default, the NoData values are transparent. This is the mask you will use to calculate slope in the next step.

It's worth noting the difference between reclassifying the cell values in a raster and merely changing its classification scheme. When you use the Symbology tab to change the number of classes displayed in a raster (five classes to nine, for instance) or to change its classification method (say, Equal Interval to Jenks), you are not changing the raster cell values. All you are changing is the scheme by which the values are symbolized. This process is commonly called classification, which can be confusing because "reclassification" is the process of changing actual cell values after their initial interpolation. When you use the 3D Analyst menu to create a slope, aspect, hillshade, or other reclassification, you are making a new raster with new cell values. When you gave the NoData cells a value of 1 and the 1 cells a value of NoData, you altered the cell values: you changed the data. Changing the classification *scheme* of a raster doesn't change the data, it just changes the way it's represented in the scene. Generally, if you change the data, you are making a new raster.

Step 9. Change the Analysis Environment

From the 3D Analyst menu, choose Options.

On the General tab, make sure that the Working Directory is set to 3DDATA\Chapter06\MyData.

Set the Analysis mask to no_lake.

On the Extent tab, set the Analysis extent to Same as Layer "sanmar."

On the Cell Size tab, also set the Analysis cell size to Same as Layer "sanmar."

Click OK to close the Options dialog.

Step 10. Calculate Slope with the No_lake Mask

Turn off all of the layers in the Table of Contents.

From the 3D Analyst menu, choose Surface Analysis, then Slope.

Set the Input surface to sanmar.

Name the Output raster sm_slope2, and save it in your 3DDATA\Chapter06\MyData folder.

Click OK.

The new slope raster is added to the scene.

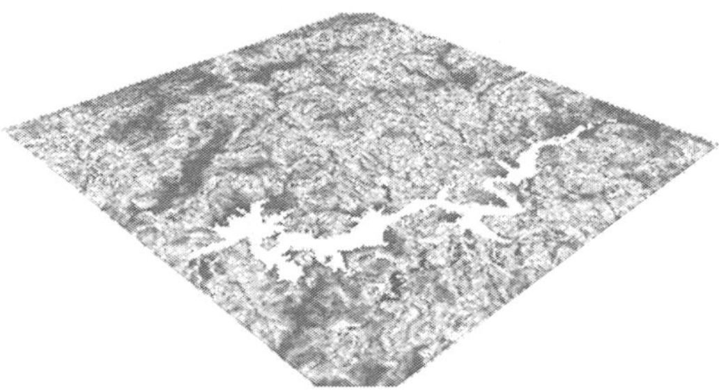

Slope was calculated wherever no_lake had cell values. The NoData values were also carried over to the new raster.

Step 11. Set Layer Properties for Sm_slope2

In the Table of Contents, double-click sm_slope2 to bring up its Layer Properties. (If ArcMap asks whether you want to compute unique values, click Yes.)

On the Base Heights tab, click "Obtain heights for layer from surface" and choose sanmar.

On the Rendering tab, check the box to shade areal features.

On the Symbology tab, click the dropdown arrow next to "Display NoData as" and choose a deep blue.

Click OK to close the dialog.

Step 12. Set Scene Properties

In the Table of Contents, double-click Scene Layers.

On the General tab, set the Vertical Exaggeration to 3.

Set the Background color to Sodalite Blue.

Click OK to close the dialog.

Step 13. Navigate the Scene

Zoom in and take a look at the new slope layer. When you're finished, give the ArcScene document a name of your choice and save it in your 3DDATA\Chapter06\MyData folder. Then close ArcScene.

Exercise 6-5

Calculate Viewshed

You learned in our earlier discussion that a viewshed raster identifies the areas of an elevation surface that can be seen from one or more observation points, and that each cell in the viewshed stores the number of observers who can see it. If you have only one observation point, each cell that can be seen from that point gets a value of 1. Cells that can't be seen from that point get a value of 0.

If more than one observer point is used, each visible cell in the viewshed stores the number of observers who can see it. Viewshed analysis assumes that the observer's field of view has no horizontal or vertical constraints, nor any distance limits.

In this exercise, you'll create viewsheds of several observation points around the city of San Luis Obispo.

Step 1. Start ArcScene and Add Data

Open ArcScene, and click the Add Data button.

Navigate to your 3DDATA\Chapter06\Data folder.

Add slocity_dem and vantage.shp to the scene.

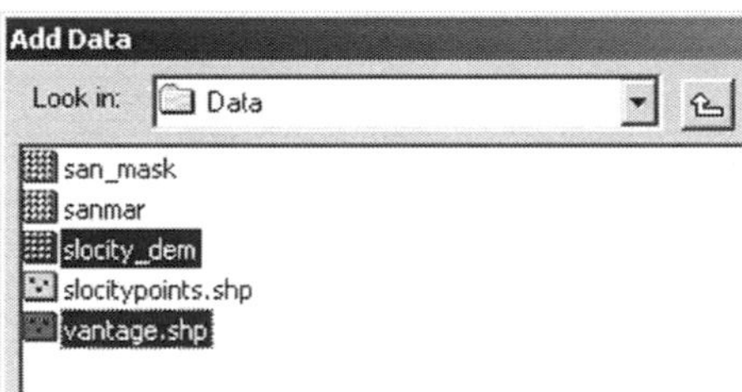

Step 2. Symbolize Slocity_dem

In the Table of Contents, double-click slocity_dem to bring up its layer properties.

On the Base Heights tab, click "Obtain Heights for Layer from Surface" and select slocity_dem.

On the Symbology tab, right-click the Color Ramp and uncheck Graphic View. Scroll down, and choose "Brown to Blue-Green Diverging, Bright."

On the Rendering tab, check the box to shade areal features.

Click OK to close the dialog.

Step 3. Set Vertical Exaggeration and Background Color

In the Table of Contents, double-click Scene Layers to bring up Scene Properties.

On the General tab, set the Vertical Exaggeration to 2, and set the Background Color to a light blue of your choice.

Click OK to close the dialog.

Step 4. Symbolize the Vantage Points

In the Table of Contents, left-click the dot symbol for the vantage shapefile.

In the Options panel of the Symbol Selector, change the symbol size to 7 and change the color to a bright red.

Click OK to close the dialog.

Step 5. Look at the Point Shapefile's Attribute Table

In the Table of Contents, right-click vantage and choose Open Attribute Table.

FID	Shape	X_coord	Y_coord	Point_ID	Elevation
4	Point	5778854.91667	2299079.16667	1	1194
1	Point	5783054.91667	2307479.16667	2	1116
3	Point	5784854.91667	2300879.16667	3	1011
0	Point	5777054.91667	2310479.16667	4	751
2	Point	5775854.91667	2304479.16667	5	446

Record: 0 Show: All Selected Records (0 out of 5 Selected.) Options

Vantage.shp has five records. Their heights are listed in the Elevation field. The Point_ID field lists the observation points in order of their elevation; 1 is the highest, 5 is the lowest. Their x and y coordinates are displayed as well.

Close the attribute table.

Step 6. Set Base Heights for the Vantage Points Layer

Double-click vantage to bring up its Layer Properties.

On the Base Heights tab, click "Use a constant value or expression to set heights for layer." Click the calculator button to bring up the Expression Builder.

You could use slocity_dem to set the base heights for the vantage points, but since the points have their own elevation values listed conveniently in their attribute table, you'll use those instead.

In the Expression Builder, select "Elevation" in the Fields box. It should appear in the Expression box, with brackets around it.

You're telling 3D Analyst to use the five values in the attribute table's Elevation field as the base heights for vantage.shp.

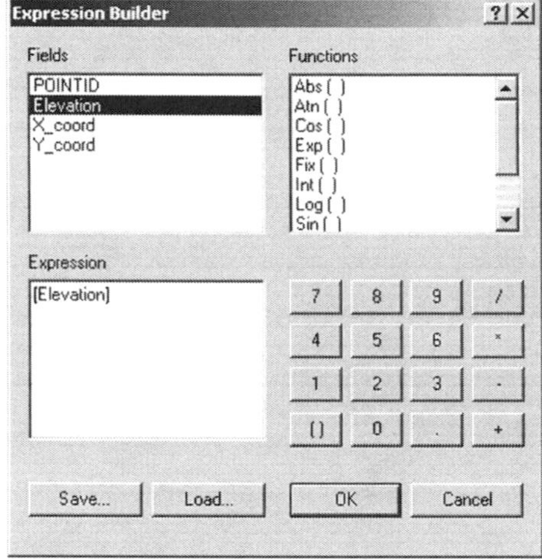

Click OK in the Expression Builder, and click OK to close the Layer Properties dialog.

The vantage points are symbolized by their elevation values in the scene, nestled in the hillsides of slocity_dem.

Step 7. Set the Working Directory

From the 3D Analyst menu, choose Options.

On the General tab, set the working directory to your 3DDATA\Chapter06\MyData folder.

Make sure that the Analysis mask is set to None.

On the Extent tab, set the Analysis extent to Same as Layer "slocity_dem."

On the Cell Size tab, set the analysis cell size to slocity_dem as well.

Click OK to close the dialog.

Step 8. Select the Highest Observation Point

In the Table of Contents, right-click vantage and choose Open Attribute Table.

Select the record for Point_ID 1 by clicking on its corresponding gray square at the left edge of the table.

FID	Shape	X_coord	Y_coord	Point_ID	Elevation
4	Point	5778854.91667	2299079.16667	1	1194
1	Point	5783054.91667	2307479.16667	2	1116
3	Point	5784854.91667	2300879.16667	3	1011
0	Point	5777054.91667	2310479.16667	4	751
2	Point	5775854.91667	2304479.16667	5	446

Record: 0 Show: All Selected Records (1 out of 5 Selected.) Options

Point_ID 1 has the highest elevation. You're going to create a viewshed raster of all the cells in slocity_dem that can be seen from that vantage point. Notice that the location has also been highlighted in the scene.

Leave the record selected, and minimize the attribute table.

Step 9. Calculate Viewshed

From the 3D Analyst menu, choose Surface Analysis, then Viewshed.

The Input surface is slocity_dem, and the Observer points layer is vantage. Only the selected point will be used in the calculation.

Call the Output raster vw_1, and save it in your 3DDATA\Chapter06\MyData folder.

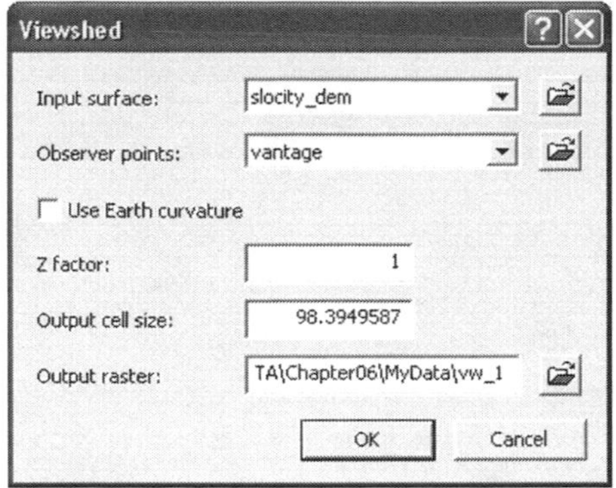

Click OK.

The viewshed from the observation point is calculated and added to the scene.

Step 10: Set Layer Properties for VW_1

In the Table of Contents, turn off slocity_dem so that you can see the new viewshed raster.

Double-click vw_1 to bring up its layer properties.

On the Base Heights tab, choose "Obtain heights for layer from surface," and choose slocity_dem.

On the Rendering tab, check the box to shade areal features.

On the Display tab, change the transparency to 0%.

Click OK to close the Layer Properties dialog.

Green cells indicate areas that can be seen from Point_ID 1; pink cells indicate areas that can't. Navigate around the surface to get an idea of Point_ID 1's perspective.

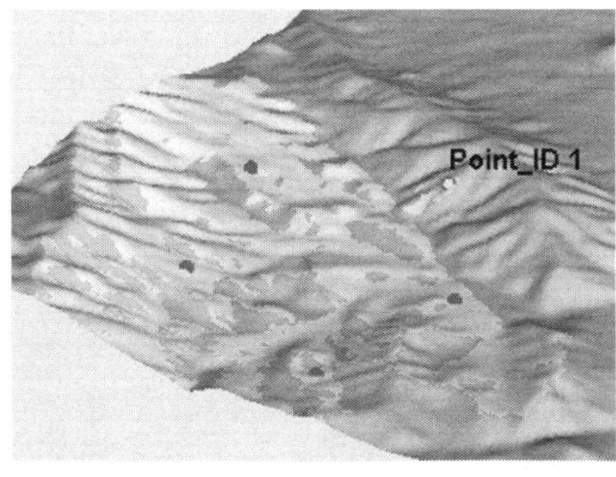

Step 11. Look at VW_1's Attribute Table

Right-click vw_1, and choose Open Attribute Table.

The cells that have a value of 1 are visible from the observation point; the cells with a value of 0 are not. The Count field tells you how many there are of each.

Close the attribute table.

In a slope or aspect calculation, small errors in cell values tend to average out in the final result, so the quality of the elevation surface doesn't matter as much. The directional compass used to measure aspect and the basic rules of geometry used to define slope are always the same. Viewshed computation, however, tends to magnify any errors in the elevation surface. The observation points from which a viewshed is calculated are always variable, so errors are more easily made and tend to compound each other. Incorrect x, y coordinates of the observer points, incorrect z values on the elevation surface, and incorrect viewshed parameters can have a snowballing effect on cell calculations. Moving an observation point a few meters can result in a very different viewshed calculation. Also, if a cell directly in front of an observer point is incorrectly given a value of 0, other cells that would have been visible will be ruled out by the value of the first incorrect cell. Finally, snapping or not snapping the output extent to the input raster can also make a considerable difference in the results.

Step 12: Clear the Selected Feature from Vantage.shp

In the Table of Contents, right-click vantage, choose Selection, and then choose Clear Selected Features.

Resize the attribute table for the vantage points shapefile. (If you closed it earlier, reopen it.)

You're going to make a new viewshed raster, using all five observation points, so make sure that none of the five records are selected.

Close the vantage attribute table.

Step 13: Calculate Viewshed for the Observer Points

From the 3D Analyst menu, choose Surface Analysis, then Viewshed.

Make sure that the Input surface is slocity_dem.

Call the Output raster vw_5 (for five observation points), and save it in your 3DDATA\Chapter06\MyData folder.

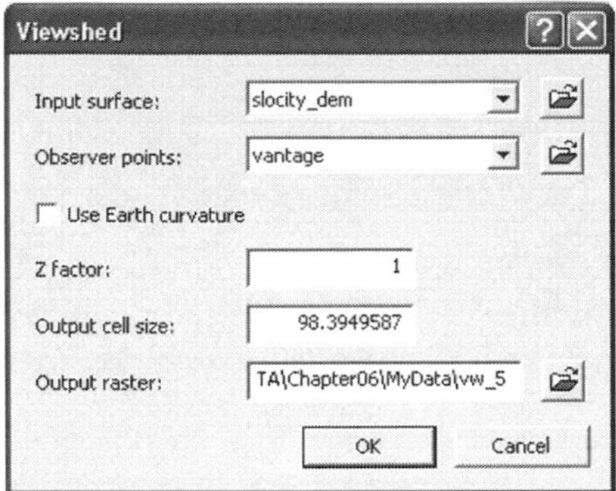

Click OK.

The process takes a few moments. When it is finished, vw_5 is added to the scene.

Step 14. Set Layer Properties for VW_5

Turn off vw_1 in the Table of Contents.

Double-click vw_5 to bring up its layer properties.

On the Base Heights tab, check "Obtain heights for layer from surface" and select slocity_dem.

On the Rendering tab, check the box to shade areal features.

On the Display tab, set the transparency to 0.

Click OK to close the dialog.

This time, the green areas can be seen from at least one of the five lookout points. Pink areas can't be seen from any point.

Step 15. Symbolize the Viewshed Layer

Double-click vw_5 to bring up its layer properties, and click the Symbology tab.

In the Show box, click Unique Values.

Right-click the color bar in the Color Scheme dropdown box, and uncheck Graphic View.

Choose Enamel from the color scheme menu.

Below the Symbol column, double-click on the color symbol next to 0, to bring up the color palette.

Choose Gray 20%.

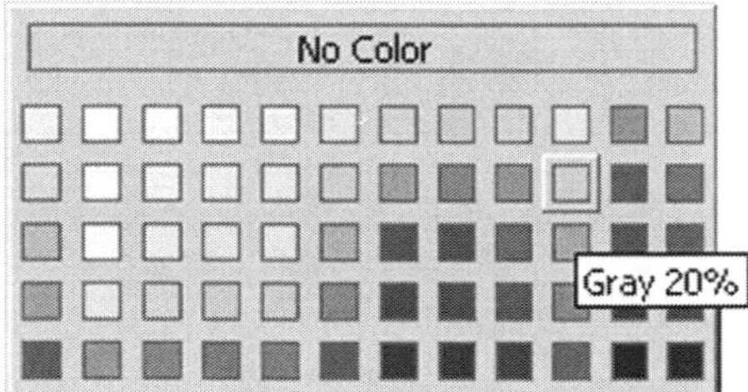

In the Label column, click on the label "0." In the highlighted text box that appears, type "Not Visible."

Press Enter.

Replace the label "1" with "1 observer," the label "2" with "2 observers," and so forth. Press Enter when you're done with each one.

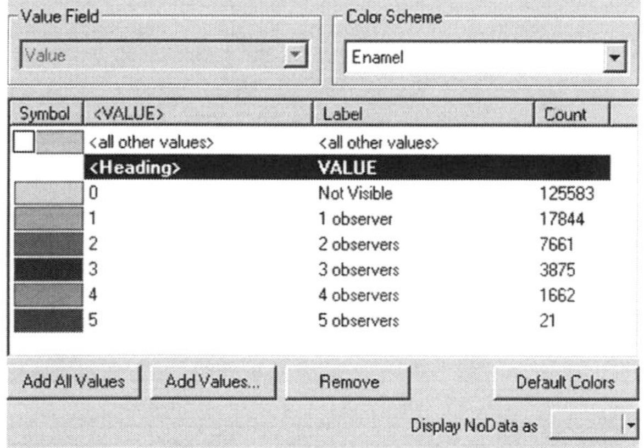

Click OK to close the dialog.

Now the new viewshed raster is symbolized by the number of observers who can see each portion of it. Note that it is not symbolized by which areas can be seen by each observer; to do that, you would have to make five viewsheds, each of the areas seen by one observer, and overlay them.

Step 16. Save the Document

Navigate the scene, and take a gander at the perspectives from different observation points. If you like, use the Set Target and Set Observer tools to test the view of the raster from the observation points themselves.

When you're finished, give the scene a name of your choice and save it in your 3DDATA\Chapter06\MyData folder. Then close ArcScene.

Challenge Exercise

Calculate Viewshed and Slope Levels for Elk Park

In this challenge exercise, you'll use ArcMap instead of ArcScene to calculate slope and viewshed for our imaginary regional park.

Step 1. Start ArcMap and Add Data

Open ArcMap, either from the icon on your desktop or Start/Programs/ArcGIS/ArcMap.

Click the Add Data button, navigate to your 3DDATA\Chapter06\Data folder, and add elk_park_bnd.shp, elk_vantage.shp, o51_dem, and o51_hshd to the map.

Step 2. Turn on the 3D Analyst Extension and Set the Analysis Environment

From the Tools menu in ArcMap, select Extensions, and make sure that the 3D Analyst extension is turned on.

If the 3D Analyst toolbar isn't visible, click the View menu, select Toolbars, and check the box next to "3D Analyst."

From the 3D Analyst menu, choose Options.

On the General tab, set the Working directory to 3DDATA\Chapter06\MyData.

Set the Analysis mask to elk_park_bnd.shp.

On the Extent and Cell Size tabs, set the analysis extent and cell size to Same as Layer "o51_dem."

Click OK.

Step 3. Create a Slope Raster of the Park

From the 3D Analyst menu, choose Surface Analysis, then Slope.

The Input surface is o51_dem, and you want to make a Percent slope raster. Call the Output elk_slope, and save it in your 3DDATA\Chapter06\MyData folder.

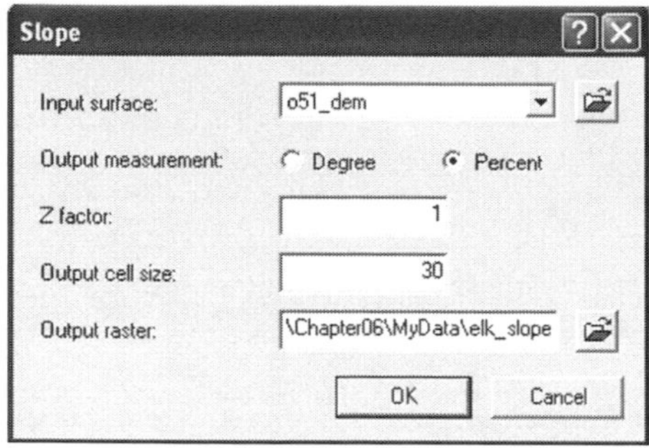

Click OK.

By setting the analysis mask to elk_park_bnd, you told 3D Analyst to only calculate slope for that area. Turn off elk_bnd.shp so that you can see the new slope raster.

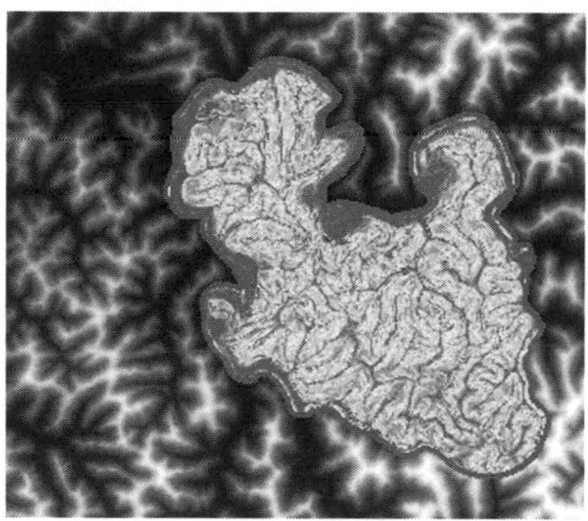

Step 4. Change the Slope Classification Values

In order to get a more useful idea of the slope in the park, you'll reduce the number of classes used to symbolize the raster.

In the Table of Contents, double-click elk_slope to bring up its layer properties. (If ArcMap asks whether you want to compute unique values, click Yes.)

On the Symbology tab, change the number of classes to four. Click the Classify button to bring up the Classification dialog, and type in Break Values of 10, 15, 20, and 186 (the highest value in this raster).

Click OK to close the Classification dialog, and OK again to close the Layer Properties dialog.

This is a very steep area—almost all of the raster cells show a slope of greater than 20 percent.

Step 5. Symbolize the Slope and Hillshade Rasters for a More Dramatic Effect

In the Table of Contents, turn off all of the layers except for elk_slope and o51_hshd. Elk_slope should be on top.

If you don't have the Effects toolbar loaded, go to the View menu in ArcMap, choose Toolbars, and check the box next to "Effects."

In the Effects toolbar, select elk_slope from the Layer dropdown menu. Click the Adjust Transparency button, and set elk_slope's transparency to about 60 percent.

The lay of the land becomes more obvious, while the colors used to symbolize slope categories provide meaningful values to work with.

Step 6. Look at the Point Shapefile's Attribute Table

Zoom to full extent.

In the Table of Contents, turn off elk_slope and o51_dem. Turn on elk_vantage and o51_hshd.

If the vantage points aren't easy to see, change the symbol size to 7 and change the color to a bright yellow.

In the Table of Contents, right-click elk_vantage and choose Open Attribute Table.

Elk_vantage.shp represents three summits in Elk Park, named after nearby towns and points of interest. The Shape field indicates that this is a 3D shapefile, with elevation values listed in the Height field.

Close the attribute table.

Step 7. Calculate Viewshed for the Park

From the 3D Analyst menu, choose Surface Analysis, then Viewshed.

The Input surface is o51_dem, and the Observer points layer is elk_vantage. All points will be used to create the viewshed, although you should feel free to experiment with viewsheds created by selecting individual points.

Call the Output raster elk_vshed, and save it in your 3DDATA\Chapter06\MyData folder.

Viewshed

Input surface:	o51_dem
Observer points:	elk_vantage
Use Earth curvature	
Z factor:	1
Output cell size:	30
Output raster:	hapter06\MyData\elk_vshed

OK Cancel

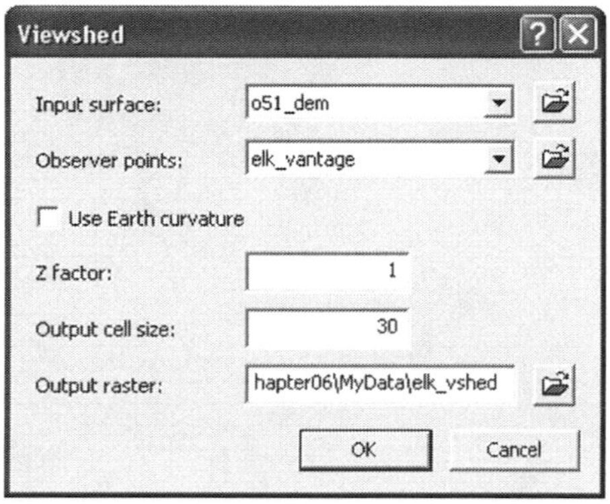

Click OK.

The viewshed from the observation points is calculated and added to the scene. Notice that the raster is automatically displayed at 20% transparency, in case you happen to have a hillshade raster that you'd like to see underneath, which we do. Also notice that even though you had the analysis mask set to elk_park_bnd, viewshed was calculated for the entire area of o51_dem. That's because the Viewshed function doesn't respect the analysis mask in the 3D Analyst Options settings.

Not very much of the park is visible from these vantage points, but most of the other summits can be seen, so these points might make good destinations for hiking trails.

Step 8. Save the Document

When you're finished, give the map a name of your choice and save it in your 3DDATA\Chapter06\MyData folder. Then close ArcMap.

CHAPTER 7

TIN Surface Models

In Chapter One you learned that a Triangulated Irregular Network is a vector data structure that represents a surface by dividing geographic space into contiguous, nonoverlapping triangles, and that each triangle node stores an x, y, and z value.

TIN Interpolation

Like a raster, a TIN is created from sample points, and their elevation values are interpolated to form a surface. A TIN is a little less complex, however; it is almost always created from elevation samples, whereas a raster can be created from samples of any phenomenon that varies continually. Further, when you create a TIN you have only one interpolation method to choose from—specifically, Delaunay triangulation.

The Delaunay method generates vector triangles from sample points so that the points become the nodes of triangles. The triangles are arranged so that a circle drawn around any triangle can contain no other nodes. This means that, collectively, the triangles are constrained to be as equiangular as possible. Like many geometric definitions, this is best explained with a picture:

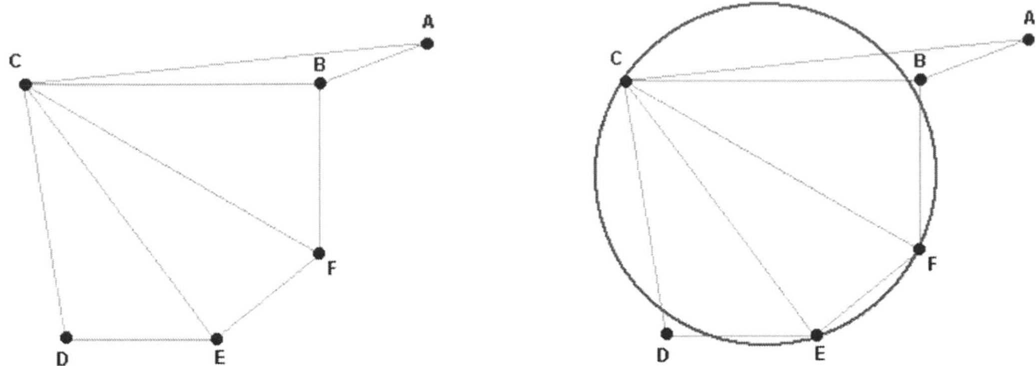

A circle drawn through the nodes of triangle CEF neither contains nor passes through any other points. This is the case for each triangle in the TIN. This keeps the triangles small and fat, which is good since long, narrow triangles are difficult for computers to process. Another feature of the Delaunay rule is that the TIN will come out the same no matter what order the points are processed in

The sample points used to create a TIN are called "mass points" because they define the bulk, or mass, of the TIN. The samples can come from point features, or from the vertices of line or polygon features. Their elevation values can come from the feature layer's attribute table, or, if it's a 3D layer, from z values in the Shape field. The sample points form the triangle nodes, and the triangulation consists of connecting the nodes by lines. Once the TIN is built, the elevation of any location on its surface can be estimated using the x, y, and z values of the bounding triangle's vertices. Each triangle in the TIN is a plane and therefore has one slope value and one aspect value, but the elevation on any part of a triangle face can be calculated from the surrounding nodes.

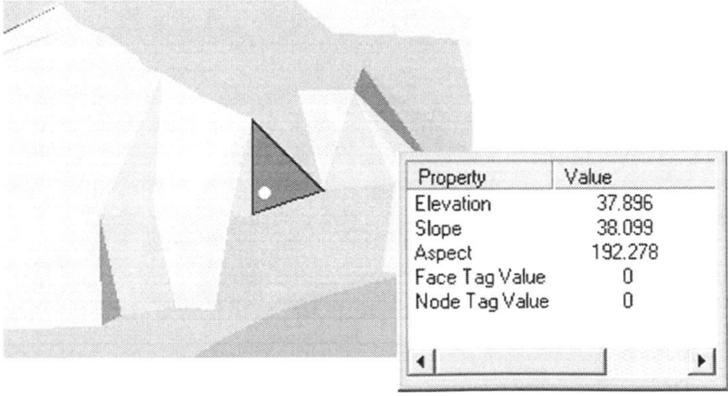

When you identify any location on the face of a TIN, the x, y, and z values of the surrounding triangle nodes are used to interpolate the elevation at that point. The nodes are also used to calculate the slope and aspect of each triangle face

Like a raster, a TIN is only as good as the initial sample points taken. Real terrain often has shapes that aren't well represented by mass point triangulation. Ridges, cliffs, lakes, and gullies are a few examples. To model these areas, you can include line and polygon features when you interpolate the TIN.

Breaklines, Replace Polygons, Clip Polygons, Erase Polygons, and Fill Polygons

There are several ways that polylines and polygons can add definition to a TIN. Breaklines are lines that tell the interpolation engine that there's a distinct change in slope on either side of the line, and that no triangle should cross it. Since locations along opposite sides of the breakline belong to different triangles, they have different slope values (as long as the surface isn't flat).

Breaklines are used to represent surface formations such as ridges, streams, dams, shorelines, and building footprints. They may or may not contain elevation values. If they do, the line vertices are used as mass points.

Top, regular mass point triangulation doesn't adequately model the river. Bottom, breaklines are used to give it definition

A replace polygon creates a flat area on a TIN surface. It stores one elevation value as an attribute. When you add the polygon, the TIN is retriangulated so that the surface area covered by the polygon has the new value.

A polygon of a reservoir is used to make the surface flat and give the banks appropriate definition

A clip polygon trims the boundary for which elevation, slope, and aspect are calculated. Clipping doesn't actually change the extent of the triangulated area, however; it just keeps TIN faces from being calculated. It does affect analysis, because the statistics for area, volume, elevation, slope, and aspect are calculated for fewer faces. Since their boundaries become triangle edges, clip polygons are often used to give a TIN a more regular shape. They may or may not contain elevation values.

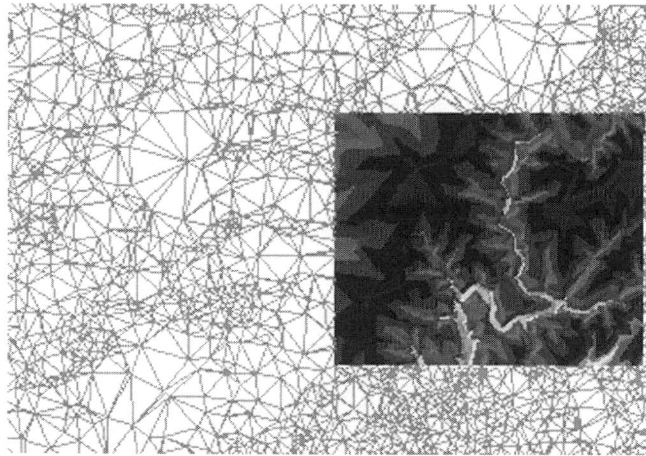

The polygon above is used to clip out a smaller zone of interpolation from the TIN. The whole TIN is still triangulated, but only the clipped area has calculations for elevation, slope, and aspect. (Normally, the additional triangles are not symbolized, but showing them here explains why, in a scene, the TIN extent will be larger than the visible clipped area.)

Erase polygons work just like clip polygons, but in reverse. They cut a hole in the TIN instead of trimming it.

Here the same polygon has been used to erase the area it overlies. Again, the whole TIN is still triangulated, but elevation, slope, and aspect are only calculated for the area outside the polygon

Fill polygons tag areas of the TIN so that the TIN can be symbolized by attributes other than its main theme of elevation, slope, or aspect. The attributes must be expressed by integer values. Fill polygons can symbolize surface features like land cover, land use, flood zones, or endangered species habitat. They cause the TIN to be retriangulated, but they don't change its elevation, slope, or aspect values.

(The term "fill polygon" seems a lot like "replace polygon" but they're quite different. A replace polygon replaces an area of a TIN with a new elevation value, such as a flat lake bottom. A fill polygon, which probably should have been called an "attribute polygon," adds a tag value of some attribute to the existing TIN without changing its shape.)

Vegetation polygons have been added as fill polygons to the TIN. Elevation values are not affected; the TIN retains its shape while areas of plant life are clearly marked

You can also assign attributes to triangle faces by specifying a tag value field from the attribute table of replace or clip polygons, and attributes can be assigned to triangle nodes by specifying a tag value from the attribute table when you add mass points. In all cases, the attribute values have to be integers. You can't use tag values with breaklines or erase polygons.

The chart below summarizes the surface feature types that can be added to a TIN, and which ones require elevation values. It comes in handy when you're trying to remember the difference between a replace, an erase, and a fill polygon.

Surface feature type	Purpose	Input layer	Elevation information	Tag Values
Mass points	Primary source of elevation values	Point, line, or polygon	Required	Optional
Breaklines	Enforce changes in slope	Line or polygon	Optional	Unavailable
Clip polygons	Define zone of interpolation	Polygon	Optional	Optional
Erase polygons	Define zone of interpolation	Polygon	Optional	Unavailable
Replace polygons	Replace with a constant elevation	Polygon	Required	Optional
Fill polygons	Assign attributes to triangles	Polygon	Optional	Purpose

TIN Symbology

In Chapter Four you learned to classify raster surfaces by elevation, slope, and aspect, and you saw that 3D Analyst provides specific color schemes to symbolize those themes. TINs are very similar, but simpler. When you reclassify a raster surface, you have to choose whether to create a new slope, elevation, or aspect raster, and 3D Analyst symbolizes it accordingly. A TIN already contains all of this information, so all you have to do is choose which of the themes you want to symbolize.

The same TIN symbolized from top to bottom by elevation, slope, and aspect

In the following exercises, you'll create a TIN from contour lines, add attribute values to it, and change its classification scheme.

Exercise 7-1

Create a TIN from Vector Features

In this exercise, you'll use a shapefile of contour lines to create a TIN of a local knoll on the outskirts of San Luis Obispo.

Step 1. Start ArcScene and Add Data

Open ArcScene, and click the Add Data button.

Navigate to your 3DDATA\Chapter07\Data folder.

Add contours.shp to the scene.

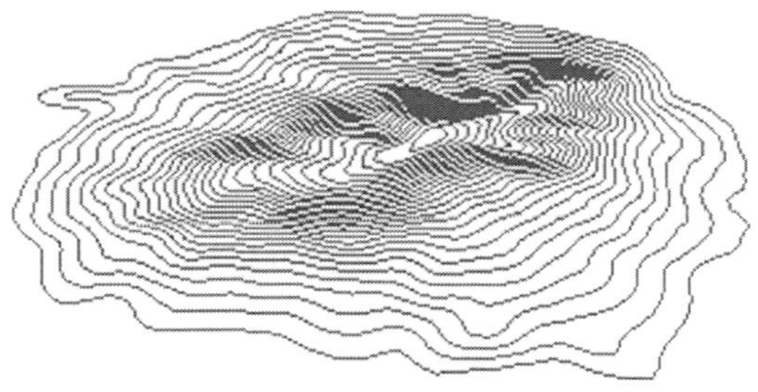

Contours is a shapefile of contour lines representing Cerro San Luis, also called San Luis Mountain and, occasionally, Cerro San Luis Obispo. It's one of a chain of small volcanic peaks that extend from Morro Bay to San Luis Obispo.

Step 2. Look at the Shapefile's Attribute Table

In the Table of Contents, right-click contours and choose Open Attribute Table.

Right-click the CONTOUR field, and choose Sort Ascending.

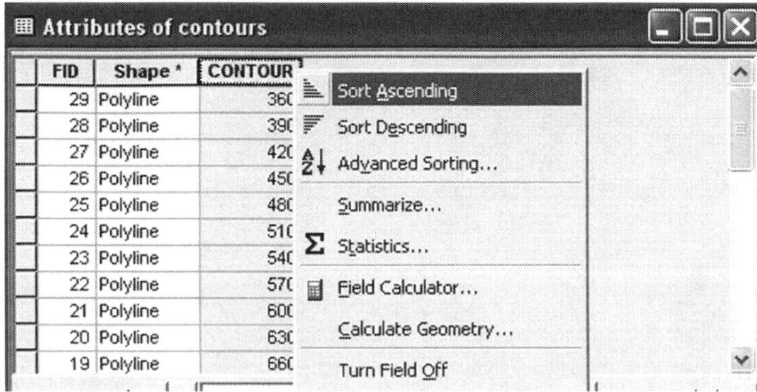

Examine the table. The contour values range from 360 to 1230 feet above sea level.

Close the table.

Step 3. Set Base Heights for Contours

In previous exercises, you used raster and TIN surfaces to set the base heights for vector layers. Since contour lines have their own elevation values, you can use the values in the CONTOUR field to set the shapefile's base heights.

In the Table of Contents, right-click contours, and choose Properties.

On the Base Heights tab, click "Use a constant value or expression . . ." and click the Calculator button.

In the Expression Builder, click CONTOUR in the Fields box. It should appear, with brackets around it, in the Expression box below.

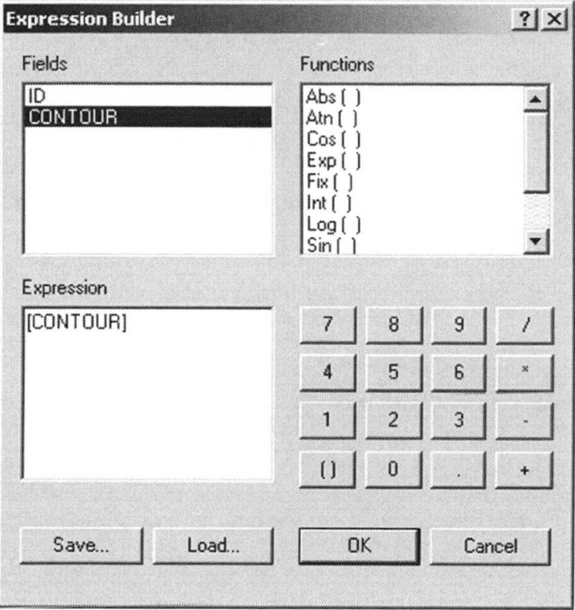

Click OK.

In the Base Heights tab, the Height panel should now be set to use the values in the CONTOUR field for base heights.

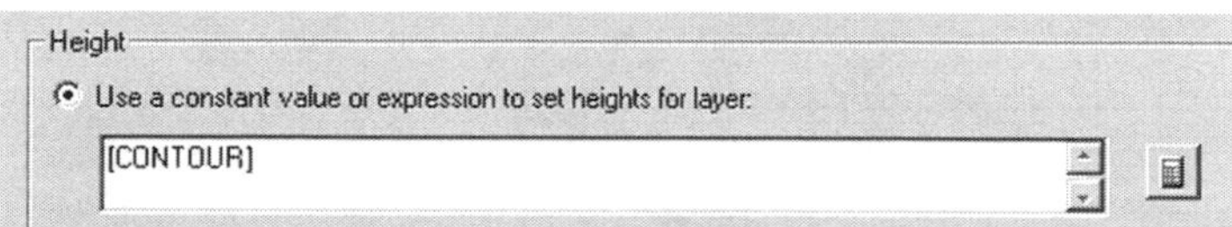

Click OK again to close the Layer Properties dialog.

Step 4. Create a TIN

From the 3D Analyst menu, choose Create/Modify TIN, then Create TIN From Features.

In the dialog under Layers, check contours.

Under Settings, set the height source to CONTOUR, and triangulate it as mass points. This tells 3D Analyst to treat all the vertices in the contour lines as a mass of x, y, z locations.

Leave the Tag Value field set to None. Name the output TIN slo_tin, and save it in your 3DDATA\Chapter07\MyData folder.

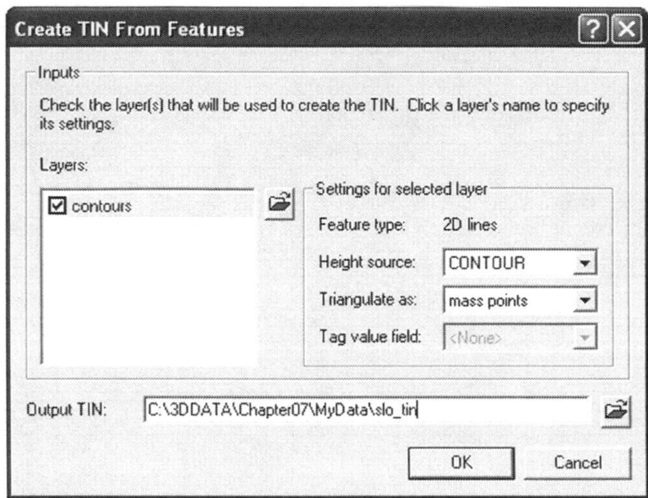

Click OK.

The TIN is added to the scene, displayed with a default symbology and with its base heights already set. Since a TIN almost always represents an elevation surface, 3D Analyst assumes that this is what you want. (A raster, as you may remember, can represent many types of phenomena besides elevation, so 3D Analyst leaves setting raster base heights up to you.)

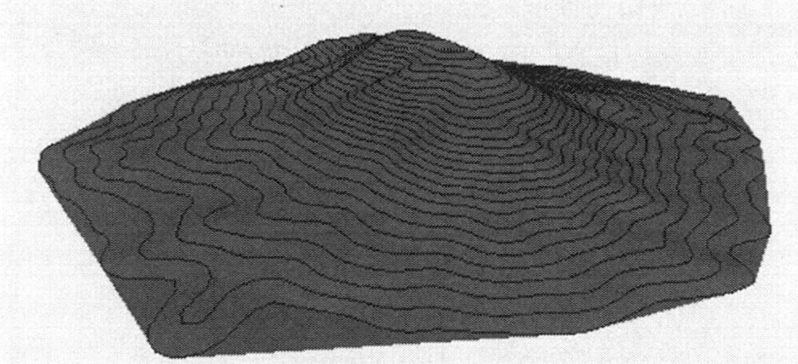

Step 5. Examine the TIN

Zoom in and navigate the surface of the TIN. Turn the contour lines on and off in the Table of Contents to get an idea of how the TIN was made from the elevation values at the vertices of each line. Also, notice that the boundary of the TIN isn't clipped to the outside contour line but instead has triangulated a more general shape around it.

Step 6. Symbolize the TIN

In the Table of Contents, double-click slo_tin to bring up its Layer Properties, and click the Symbology tab. In the Show box, uncheck Faces. Click the Add button, and choose "Face elevation with graduated color ramp." Click Add in the Add Renderer, then Dismiss.

Right-click the color ramp in the dropdown list, and uncheck Graphic View.

Scroll up and choose the Blue-Green Bright color ramp.

Click OK.

Step 7. Add the Hill's Boundary Polygon

Click the Add Data button.

Navigate to your 3DDATA\Chapter07\Data folder, and add slo_hill_bnd.shp to the scene.

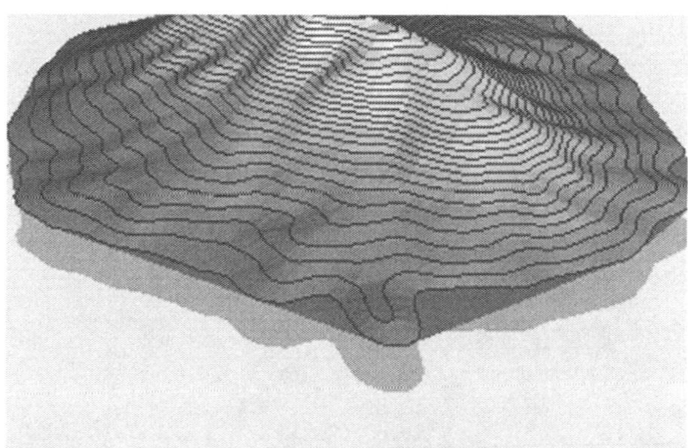

Step 8. Clip the TIN to the Contour Boundary

From the 3D Analyst menu, choose Create/Modify TIN, and then Add Features to TIN.

The input TIN is set to slo_tin.

In the Layers panel, check the box next to slo_hill_bnd.

Set the Height Source to None, and triangulate the layer as a hard clip.

Leave the tag value set to None, and make sure that the changes will be made to the existing input TIN.

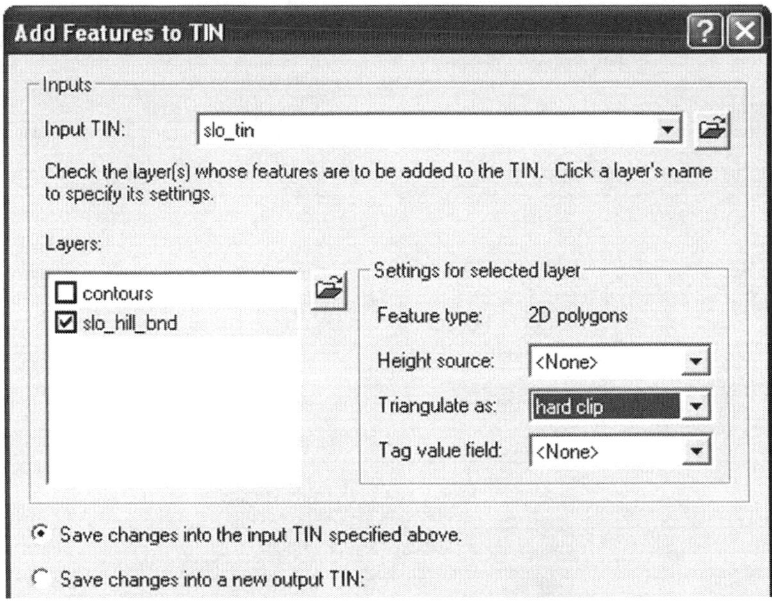

Click OK. After a few moments, the TIN is clipped to the shape of the polygon.

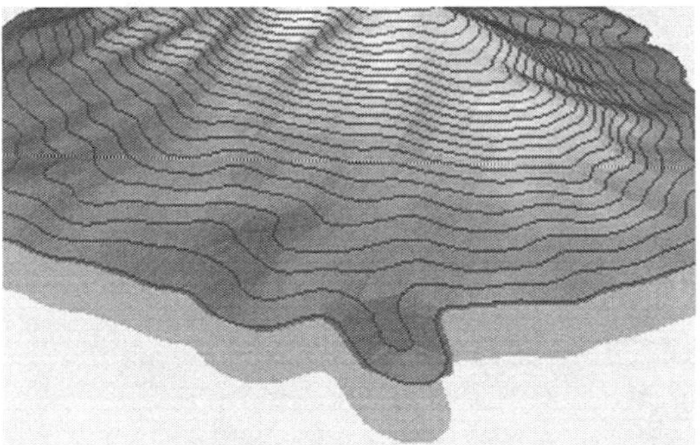

Turn off contours and slo_hill_bnd in the Table of Contents.

When you first created the TIN, you could have triangulated the mass points and clipped the TIN's boundary in one step instead of two. The Create TIN from Features dialog lets you set the parameters for each layer involved and then does the operation all at once. To keep things from getting too confusing in this first illustration, though, it made more sense to do the operations separately.

Step 9. Navigate the Hill and Use the ID Tool

Use the Zoom In/Out and Navigate tools to take a look at the TIN you've created. Click the ID button and then click anywhere on the TIN.

The ID tool tells you, for any location that you click on, the TIN's elevation, slope, and aspect. A TIN face has only one slope value and aspect value, but the elevation varies along the face, and 3D Analyst interpolates this value using the face's three nodes. If you zoom in far enough to click on several different locations within a single face, you'll get the same slope and aspect value each time, but different elevation values.

Step 10. Close ArcScene

When you're finished using the ID tool, give the scene a name of your choice and save it in your 3DDATA\Chapter07\MyData folder.

Exercise 7-2

Add Polygon Attribute Values to a TIN

The peak of Cerro San Luis divides the Stenner Lake watershed from the Laguna Lake watershed in San Luis Obispo County. In this exercise you'll symbolize the TIN of Cerro San Luis with polygons of the watershed boundaries.

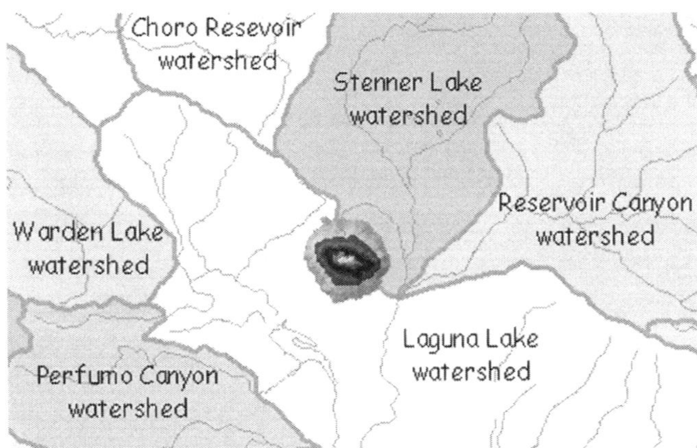

Step 1. Start ArcScene and Add Data

Open ArcScene and click the Add Data button.

Navigate to your 3DDATA\Chapter07\Data folder, and add shed_tags.shp, slo_streams.shp, and slo_tin_bkup to the scene.

Step 2. Examine the Data

In the Table of Contents, right-click slo_tin_bkup and choose Zoom to Layer. This is a copy of the TIN you made in the last exercise.

Shed_tags lies underneath the TIN. It's a polygon layer of the Stenner and Laguna Lake watersheds, clipped to the extent of Cerro San Luis.

Turn off slo_tin_bkup in the Table of Contents so you can see the polygons. Zoom into the split between them. It's a very small distance, only about two meters. The split indicates the ridgeline that separates the two watersheds.

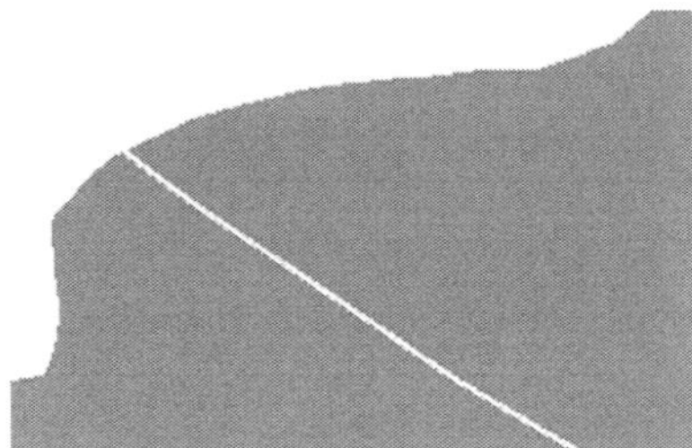

Open the shed_tags attribute table. The table lists the acreage and names of the watersheds. It also includes a Tag field with a code value for each watershed, which you'll use when you incorporate the polygons into the TIN. (Tag values have to be integers, so you can't just tag the TIN with the names of the watersheds.)

Attributes of shed_tags

FID	Shape*	NAME	ACRES	TAG
0	Polygon	Stenner Lake	7259	100
1	Polygon	Laguna Lake	18172	200

Record: 0 Show: All Selected Records (0 out of 2 Selected.)

Close the attribute table.

Step 3. Set Base Heights for the Polygons

Turn slo_tin_bkup on again.

Zoom to the extent of shed_tags, and open its Layer Properties.

On the Base Heights tab, click "Obtain Heights for layer from surface" and choose slo_tin_bkup as the source.

In the Offset panel, type a value of 150 so that the polygons won't compete with the TIN for the same 3D space.

Click OK.

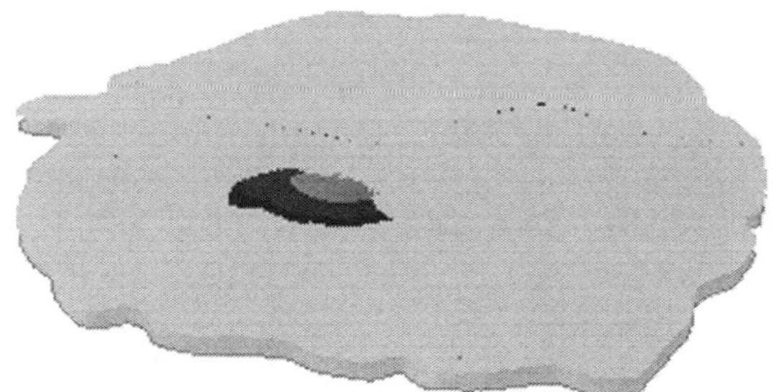

The watershed polygons are displayed with the TIN's base heights and an offset of 150 feet, but you can still see the TIN peeking through. You could turn the TIN off, or give it a lower drawing priority, but the polygons still wouldn't have the TIN's quality of detail. That's because base heights for vector features can only be set for their vertices, which works fine for points or lines but not so well for the vertex-free area inside a polygon.

Step 4. Add the Polygon Tag Attributes to the TIN

Zoom back to the extent of slo_tin_bkup and shed_tags.

From the 3D Analyst menu, choose Create/Modify TIN, then Add Features to TIN.

In the Layers panel, check the box next to shed_tags.

Set the height source to None, triangulate it as a soft value fill, and set the tag value to TAG.

Save it as a new output TIN in your 3DDATA\Chapter07\MyData folder, and call it "slo_tin_shed."

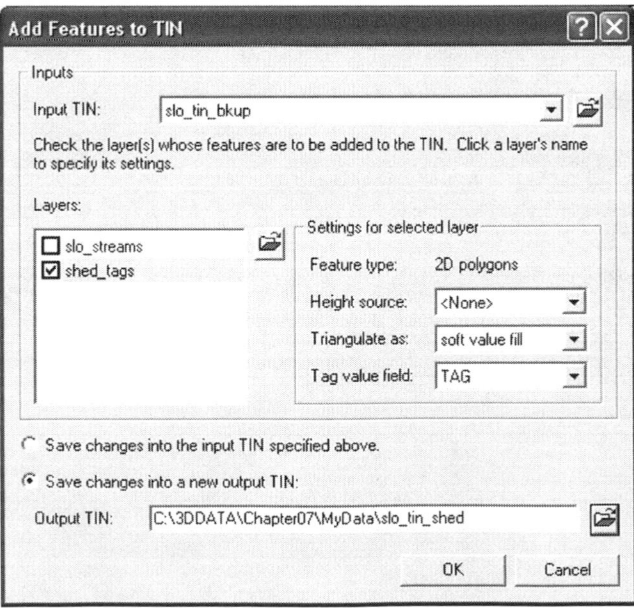

Click OK.

After a while, the new TIN is added to the scene. It looks just like slo_tin_bkup but with a line running down the middle.

Step 5. Symbolize the TIN by the Watershed Polygons

Remove shed_tags and slo_tin_bkup from the Table of Contents.

Right-click slo_tin_shed and bring up its Layer Properties.

On the Symbology tab, uncheck Faces and Edge Types in the Show box.

Click the Add button and choose "Face tag value grouped with unique symbol."

Click Add, and then Dismiss.

The Face tag values are added to the Show box, and should be highlighted.

Next, uncheck the box next to "all other values" in the main Symbol panel.

Click Add All Values.

Three attribute symbols are added: 100 for the Stenner Lake watershed, 200 for the Laguna Lake watershed, and 0 for the rest—essentially a NoData value.

Double-click the symbol next to "100" to bring up the Symbol Selector. Click the Fill Color dropdown arrow, and choose Rhodolite Rose.

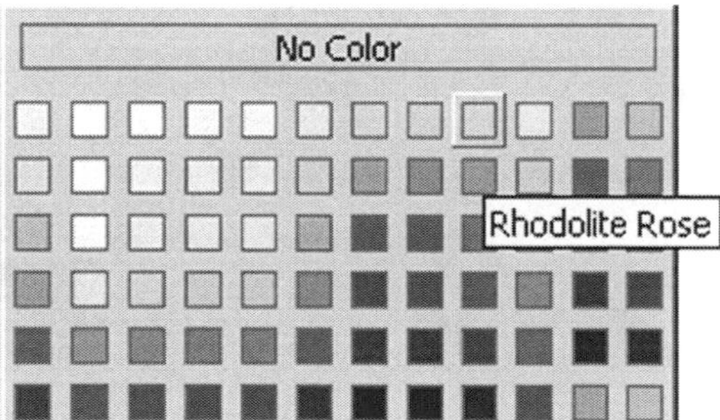

Click OK to close the Symbol Selector.

Double-click the symbol next to "200," and change its fill color to Ultra Blue.

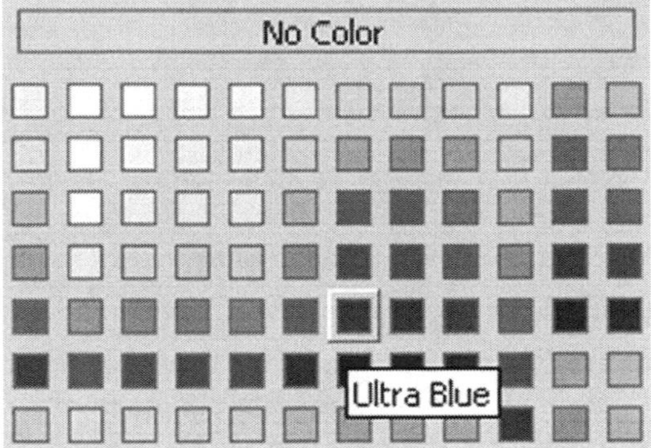

Click OK to close the Symbol Selector.

When you're finished, the Symbology tab should look like this:

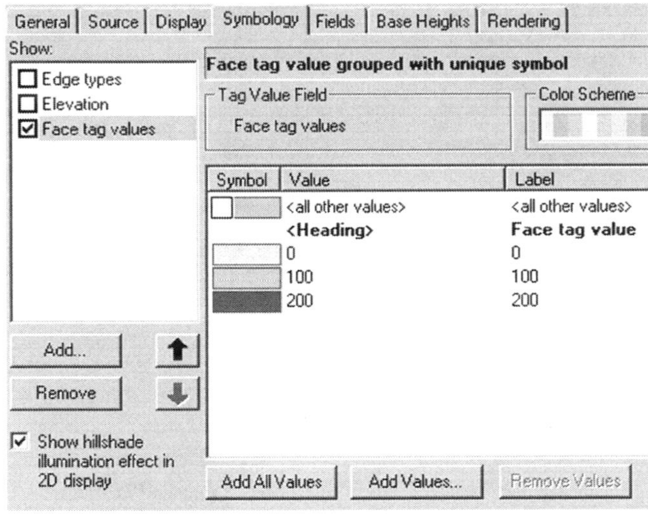

Click OK.

Cerro San Luis is symbolized by the Laguna and Stenner watersheds.

Step 6. Set Vertical Exaggeration

In the Table of Contents, double-click Scene Layers to bring up the Scene Properties dialog. On the General tab, set the vertical exaggeration to 1.5.

Click OK.

Zoom in and navigate around the mountain. Notice that all the detail of the TIN is retained in the watershed polygons.

Step 7. Set Base Heights for the Stream Shapefile

In the Table of Contents, left-click the line symbol under slo_streams to bring up the Symbol Selector.

Change the line color to Sodalite Blue, and set the line width to 2.

Click OK in the Symbol Selector.

In the Table of Contents, double-click slo_streams to bring up its Layer Properties. On the Base Heights tab, click "Obtain layer heights from surface" and choose slo_tin_shed as the source.

Click OK.

Two things happen: the streams are raised to the elevation of the TIN, but they're also clipped to the extent of the TIN. The only visible stream left is Old Garden Creek, which starts on the northern slope of Cerro San Luis.

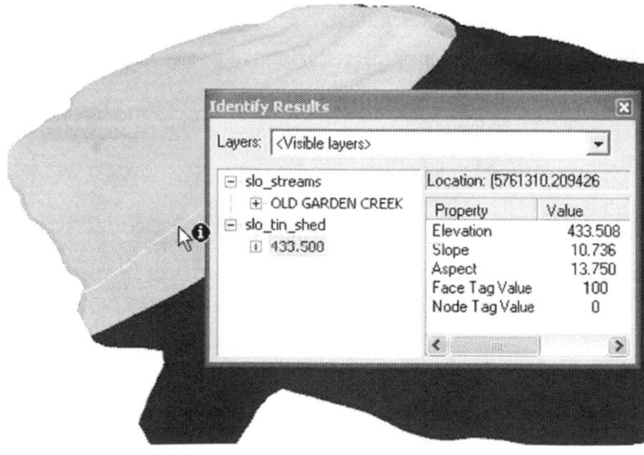

There are ways around this, the easiest being to add the streams again as a separate layer, this time not setting their base heights.

Step 8. Add the Local Stream and Watershed Layer Files

Click the Add Data button, navigate to your 3DDATA\Chapter07\Data folder and add local_sheds_lyr.lyr and slo_streams_lyr.lyr. (For version 9.1 or 9.2, use the corresponding "v91" layer files.)

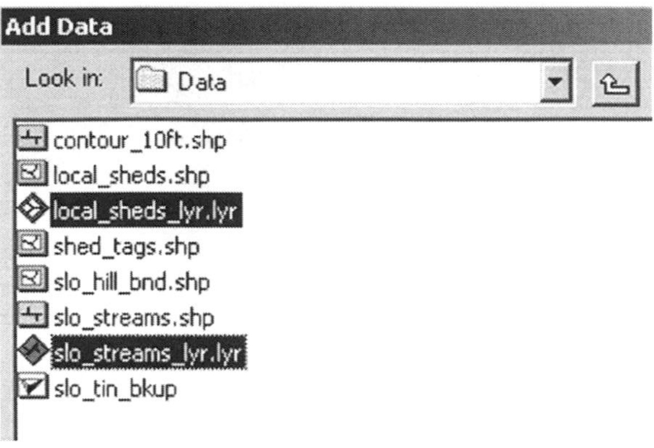

Zoom to Full Extent, and then zoom to the extent of slo_tin_shed. Take a turn around the data.

Slo_streams_lyr and local_sheds_lyr have been symbolized to match the streams and watersheds on Cerro San Luis. Their base heights are set to 0 so that they won't be clipped to any extent other than their own. However, because the TIN starts at an elevation of 360 feet above sea level, it floats 360 feet above the layer files.

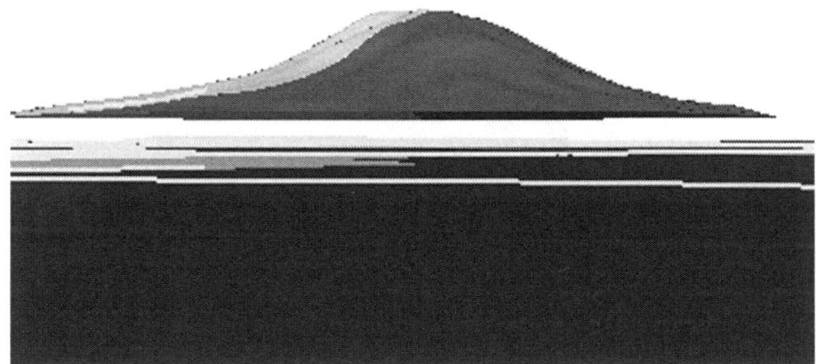

Step 9. Offset the Stream and Watershed Layer Files

Bring up the Layer Properties for slo_streams_lyr.

On the Base Heights tab, set an offset of 360.

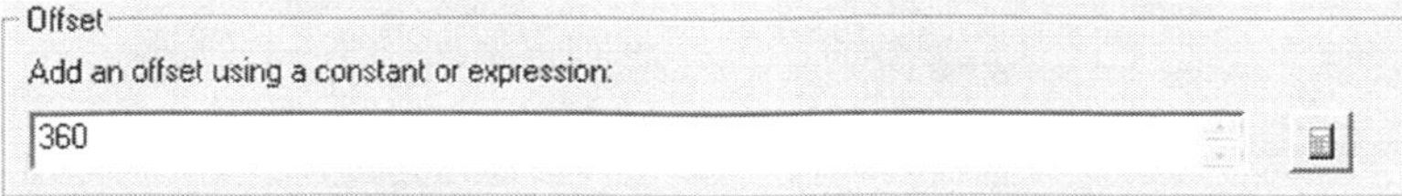

Click OK.

Give local_sheds_lyr the same offset.

Navigate around the scene. The TIN of Cerro San Luis is symbolized by its watershed boundaries, and the surrounding streams and watersheds are displayed at the correct elevation.

Artist's tip: to disguise the NoData values peeking out along the edge of slo_tin_shed, click the symbol next to the "0" value under slo_tin_shed in the Table of Contents, and change it to Ultra Blue.

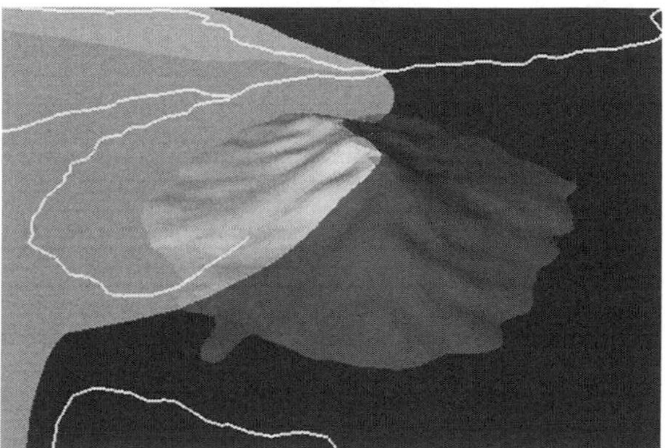

Step 10. Close ArcScene

When you're finished checking out the data, close ArcScene. Give the scene a name of your choice and save it in your 3DDATA\Chapter07\MyData folder.

Exercise 7-3

Change TIN Symbology and Classification

The cells in a raster can only hold one type of value, so when you create a surface raster, you have to choose the type of phenomena that it will represent: elevation, slope, or aspect. When you make a TIN, all of that information is included, and you can represent elevation, slope, or aspect just by changing the TIN's symbology. In this exercise you'll change the ways your TIN of Cerro San Luis is symbolized and classified.

Step 1. Start ArcScene and Add Data

Open ArcScene and click the Add Data button.

Navigate to your 3DDATA\Chapter07\Data folder, and add slo_tin_bkup to the scene.

The TIN is displayed by Edge types, and by Faces with the same symbol.

Step 2. Change Vertical Exaggeration and Background Color

In the Table of Contents, double-click Scene Layers. On the General tab, set the vertical exaggeration to 1.5, and change the background color to a blue of your choice.

Click OK.

Step 3. Get to Know the TIN's Edge Types

In the Table of Contents, double-click slo_tin_bkup to bring up its Layer Properties. Click the Symbology tab.

In the Show box, uncheck Faces.

Select Edge types, and under the Symbol panel click Add All Values.

Three line types show up: 0, 2, and 3, along with their descriptions.

Double-click the line next to 0 (Regular Edge). When the Symbol Selector pops up, choose a thin, medium-gray line symbol.

Click OK. Do the same for the other three symbols, but choose a fat red line for 2 (Hard Edge) and a thin white line for 3 (Outside Edge).

Line Type 1 is missing because that is reserved for Soft Edge, and there aren't any in this TIN. Soft edges are often used to delineate study areas instead of geographical boundaries. The TIN you made in Exercise 2, slo_tin_shed, has soft edges defining the watershed polygons, because you triangulated them as a "soft value fill."

When you're done, the Symbol panel should look something like this:

Symbol	Value	Label	Count
	<all other values>	<all other values>	0
	<Heading>	**Edge type**	**26826**
	0	Regular Edge	24433
	2	Hard Edge	1225
	3	Outside Edge	1168

Click OK.

Zoom in and take a look. The hard edge is the line made by the outside contour line, which you used when you clipped the TIN in Exercise 1. The outside edges make up the area beyond the contour line, and the regular edges are the bulk of the TIN. Looking at the edges gives you a better understanding of how the TIN is triangulated. (Soft and hard edge types are often referred to as soft and hard breaklines in the ArcGIS Help.)

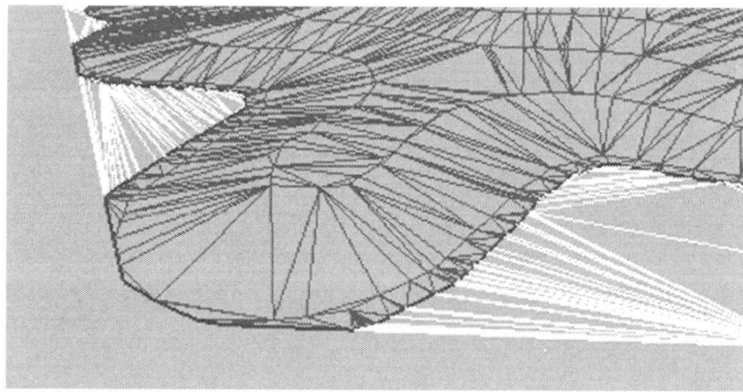

Step 4. Add Slope, Aspect, and Elevation Symbol Schemes

Bring up the TIN's layer properties, and click the Symbology tab, if necessary.

In the Show box, select Faces, and click Remove. Do the same for Edge Types.

Click Add button.

Select "Face slope with graduated color ramp" and click the other Add button.

Also add "Face aspect with graduated color ramp," and "Face elevation with graduated color ramp."

Click Dismiss.

Uncheck the Elevation and Aspect schemes in the Show box, and select Slope to highlight it.

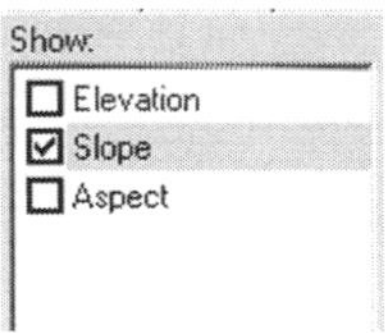

Click OK.

3D Analyst symbolizes each face of the TIN by its degree of slope. The ranges are classified into nine categories by default, but you can change that. You can also change the color scheme, and individual colors within it.

Step 5. Change Slope Classification

Open slo_tin_bkup's Layer Properties. On the Symbology tab, highlight Slope in the Show box. (3D Analyst highlights the top entry in the Show box each time you reopen the Layer Properties. If you like, you can use the arrow buttons to put Slope at the top of the Show box.)

Click the Classify button.

Change the number of classes to 4. In the Break Values panel, change the breaks to 5, 10, 20, and 90.

Click OK to close the Layer Properties dialog.

Now Cerro San Luis is symbolized by four slope value ranges.

If you like, experiment with the slope values by using a Natural Breaks classification and increasing the number of classes. Just remember that Slope won't be automatically highlighted in the Show box unless you move it to the top.

Step 6. Symbolize the TIN by Aspect

Open TIN's Layer Properties dialog and click the Symbology tab again. In the Show box, uncheck Slope and check Aspect.

Click OK.

The default color scheme has 8 classes for the 8 major compass directions, plus a gray symbol for flat areas. The North symbol is shown twice.

Being almost circular, Cerro San Luis lends itself well to being symbolized by all eight directions, so you won't really improve upon it by changing its default settings. In many cases, however, you might want to simplify the color scheme to show that most areas face one direction or another.

Step 7. Look at the TIN in Orthographic View

From the View menu, choose View Settings.

Change the scene to Orthographic. Move the dialog out of the way so you can see the scene.

Looking at your data in orthographic view is a good way to be sure of your orientation. Orthographic view mimics the perspective of ArcMap, with North always at the top.

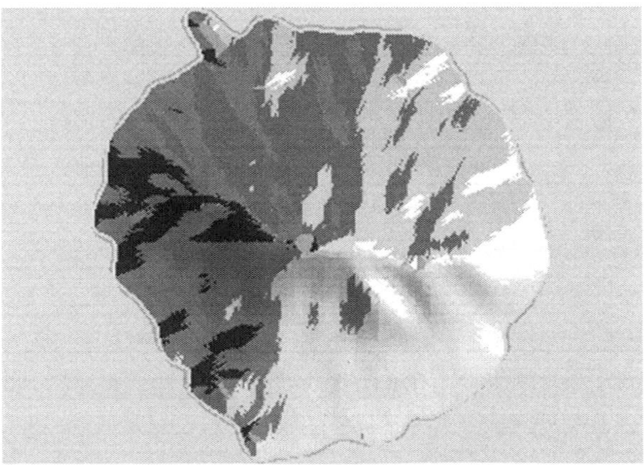

You can see that the ridgeline runs roughly East-West. If you change the aspect color scheme, you can emphasize this.

Step 8. Change Aspect Classification

In the Table of Contents, right-click the Northeast symbol.

Change the color to Mars Red.

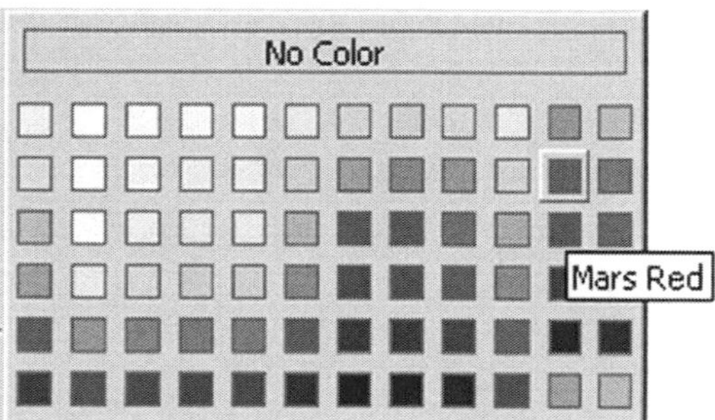

Also change the Northwest symbol to Mars Red.

Select the symbol for South, and change it to Yogo Blue.

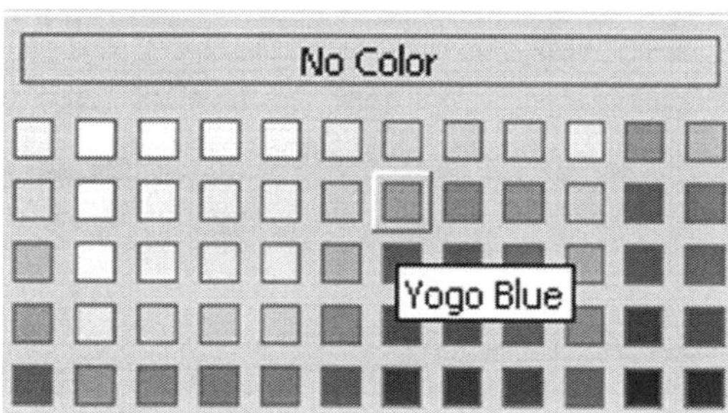

Do the same for the Southwest and Southeast Symbols.

Now it's more obvious that the ridge divides the hilltop into northern and southern directions.

In View Settings, change the scene back to Perspective view, and close the dialog.

Step 9. Experiment with Elevation Classification

Double-click slo_tin_bkup in the Table of Contents.

On the Symbology tab, uncheck Aspect in the Show box, and check Elevation.

Change the number of classes, and pick a color scheme that appeals to you. Experiment with the individual colors for each symbol, if you like.

You can create some pretty cool (and even informative) surface features by changing a TIN's elevation, slope, and aspect symbology.

Step 10. Close ArcScene

Feel free to display the TIN with the other types of symbology listed in the "Add Renderer" dialog. When you're finished experimenting, give the scene a name of your choice and save it in your 3DDATA\Chapter07\MyData folder.

Challenge Exercise

Create a TIN of Elk Park

In this exercise you'll use ArcMap to create a TIN of Elk Park from contour lines. You'll also refine the TIN by adding polygons of the park boundary and summit areas.

Step 1. Start ArcMap and Add Data

Open ArcMap and click the Add Data button.

Navigate to your 3DDATA\Chapter07\Data folder, and add elk_contour.shp, elk_park_bnd.shp, elk_vantage.shp, and summits.shp.

Step 2. Symbolize the Layers

In the Table of Contents, arrange the layers so that elk_vantage is on top, followed by elk_contour, summits, and elk_park_bnd.

The three polygons in the summits shapefile were created using the contour lines around the three points of elk_vantage.shp. The idea is to create a plateau around each point that can be used to make the summits more visible on the TIN and easier to find in later applications, such as Google Earth.

Symbolize the points of elk_vantage so that you can see them easily, but so they don't obscure the summit polygons.

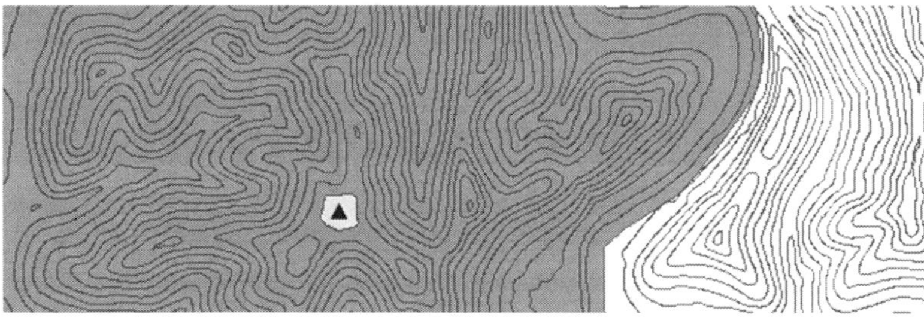

Take a look at the rest of the layers. Elk_contour is a shapefile of 50-foot contour lines created from a clipped version of o51_dem. Elk_park_bnd should be pretty familiar to you by now.

Step 3. Set the Analysis Environment

From the Tools menu in ArcMap, select Extensions, and make sure that the 3D Analyst extension is turned on.

If the 3D Analyst toolbar isn't visible, click the View menu, select Toolbars, and check the box next to "3D Analyst."

From the 3D Analyst menu, choose Options.

On the General tab, set the Working directory to 3DDATA\Chapter07\MyData, and set the Analysis mask to None.

On the Extent tab, set the Analysis extent to "Intersection of Inputs." On the Cell Size tab, the Analysis cell size should be "Maximum of Inputs."

Step 4. Create the TIN

From the 3D Analyst menu, choose Create/Modify TIN, then Create TIN From Features.

In the dialog under Layers, check elk_contour.

Under Settings, set the height source to CONTOUR, and triangulate it as mass points.

Leave the Tag Value field set to None. Name the output TIN elk_tin and save it in your 3DDATA\Chapter07\MyData folder.

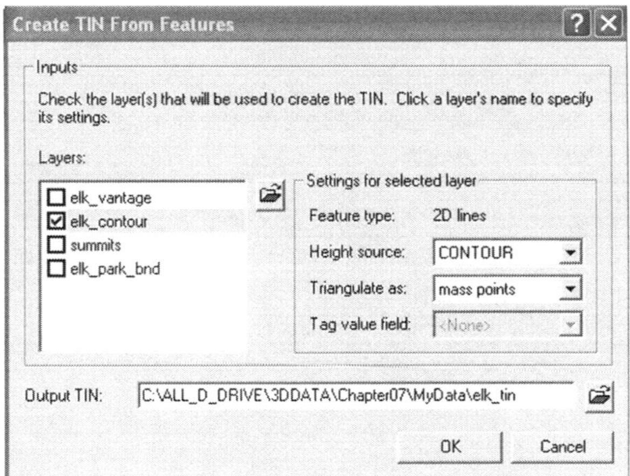

Click OK.

After a while, the new TIN appears. It's much larger than the TINs you worked with previously—it contains nearly 900,000 triangles.

Arrange the layers in the Table of contents so that elk_tin is below elk_contour and elk_vantage. You can see that the contours effectively act as break lines during the TIN's triangulation.

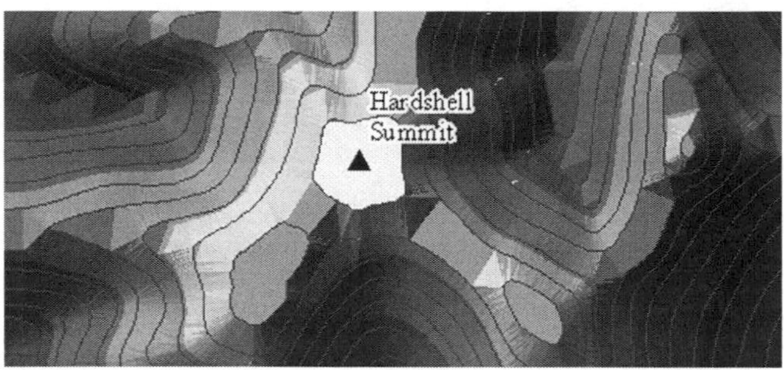

Step 5. Add Elk_park_bnd and the Summit Polygons to the TIN

Zoom to Full Extent.

From the 3D Analyst menu, choose Create/Modify TIN, then Add Features to TIN.

The input TIN is elk_tin.

In the dialog under Layers, make sure elk_contour and elk_vantage are unchecked.

Check the box next to summits.

While summits is selected, set the Height source to None, triangulate it as a soft value fill, and set the Tag value field to CONTOUR.

Now check elk_park_bnd under Layers.

While elk_park_bnd is selected, set the Height source to None, triangulate it as a hard clip, and set the Tag value field to None.

Finally, save the changes in a new output TIN, called elk_tin2, and put it in your 3DDATA\Chapter07\ MyData folder.

Click OK.

The new TIN is added to the map, noticeably clipped to the extent of Elk Park. What may be less noticeable is that the three summit polygons have been added to the TIN as a tag values.

Step 6. Symbolize Elk_tin2

In the Table of Contents, turn off all of the layers except elk_tin2.

Double-click elk_tin2 to bring up its Layer Properties.

On the Symbology tab, remove both Edge Types and Elevation from the Show box. Click the Add button, and choose "Face tag value grouped with unique symbol." Click Add and Dismiss in the Add Renderer.

Next, uncheck the box next to "all other values" in the main Symbol panel.

Click Add All Values.

Two attribute symbols are added: 1400 for the tag values assigned by the summit polygons, and 0 for the rest of the TIN.

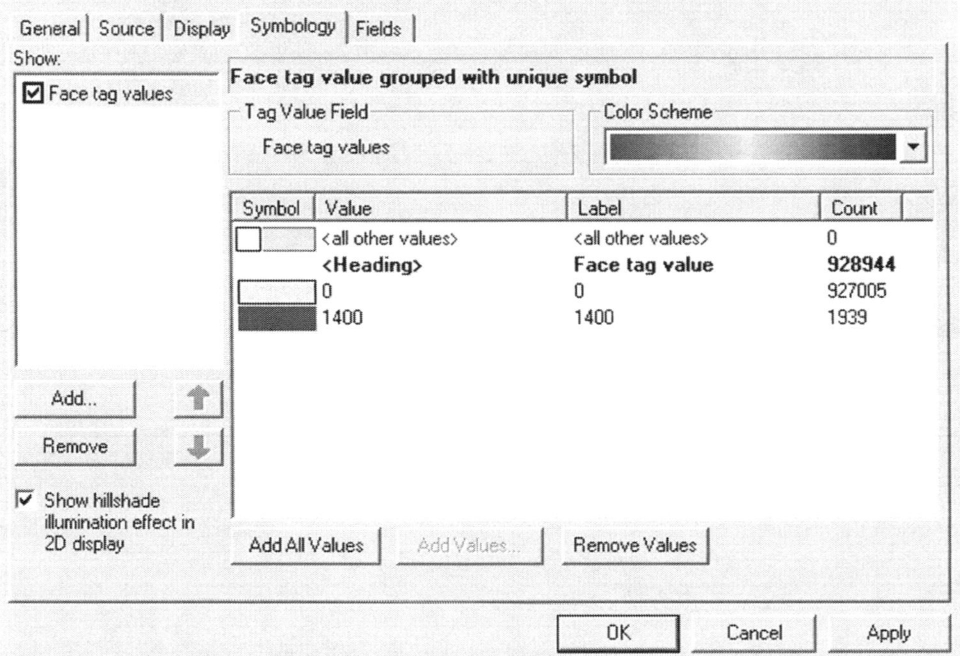

Choose colors that aren't too garish for the tag values; perhaps a light green for the bulk of the TIN faces and a dark green for the summits.

Click OK to dismiss the Layer Properties dialog, and take a look at the map.

The values of "1400" for the summit polygons are derived from their contour line height values, but they aren't being used as heights for the TIN. They're serving only as tag identifiers, just as the tag values in the watershed polygons that you used in Exercise 2 were unique identifiers. We could have given each of these summits their own integer value—say, 100, 200, and 300—but since they're all plateaus at the same elevation it made sense to leave them as one category of tag value. The rest of the triangles in the TIN have a tag value of 0.

If you like, you can experiment with the edge, face, and elevation types in the Symbology tab to show the elevation values of the TIN while also highlighting the summit areas. See if you can arrive at the following effect, though not in black and white, of course:

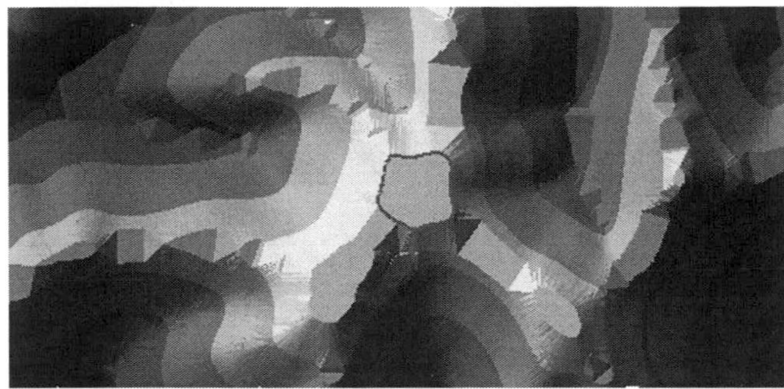

Hint: there are five elevation levels shown. Face tag values only render the summits, and edge types showcase the soft edges of the summit fill. The key to getting these to show up is the order they're listed in the Show box, in the Symbology tab.

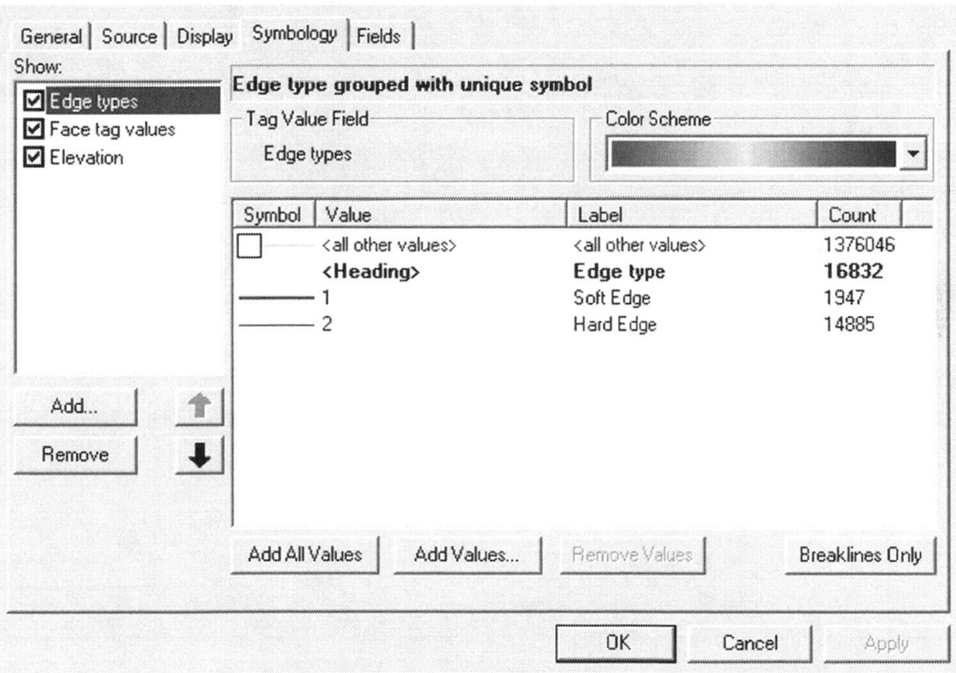

When you're done experimenting, give the ArcMap document a name of your choice and save it in your 3DDATA\Chapter07\MyData folder. Also feel free to open your new TINs up in ArcScene and experiment with the symbology there.

CHAPTER **8**

Terrain Surface Models

Now that you understand the TIN data structure, it's time to tackle the terrain surface model. The terrain is new to ArcGIS 9.2 and 9.3, and can be described as a multi-level TIN created from hundreds of thousands of mass points, organized into pyramid levels of detail so that at large distances, a coarser resolution is used, and at close distances more detail is revealed. Thus, terrains can be added to ArcGlobe and viewed seamlessly at all scales, without using all your RAM or breaking your graphics card.

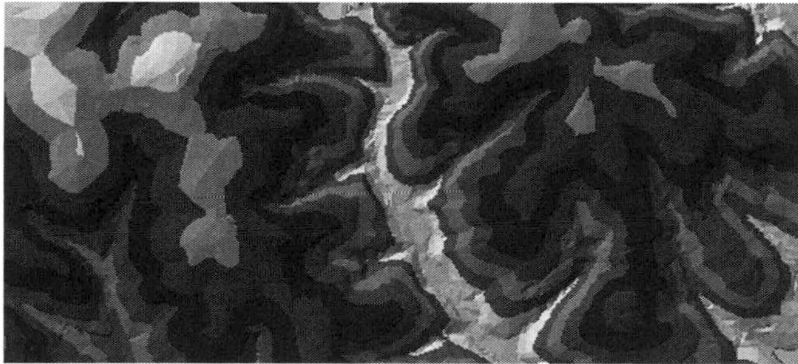

A terrain at a scale of 1:15,000

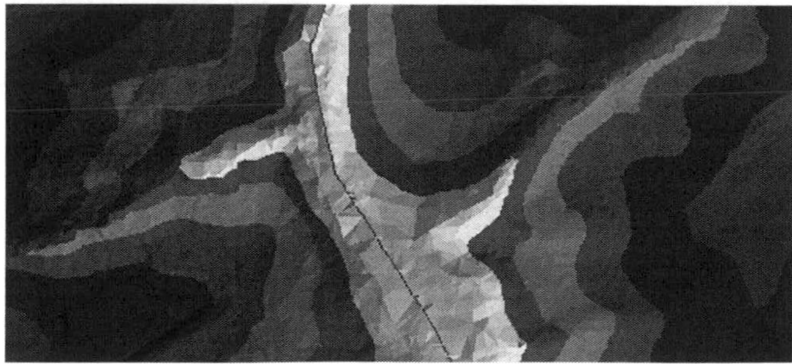

The same terrain at a scale of 1:5,000

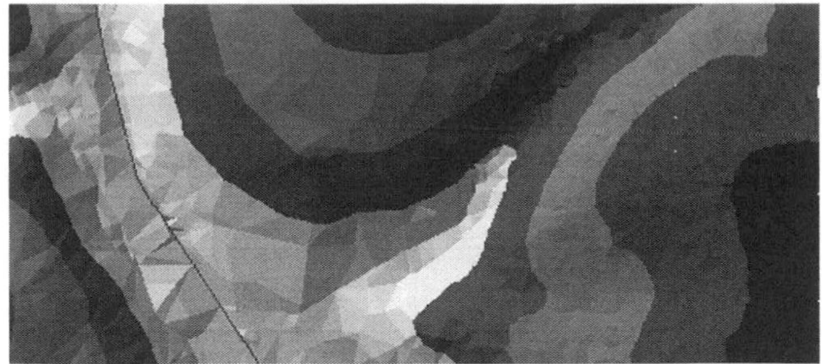

The same terrain at a scale of 1:2,000

One way that Terrains differ from TINs is that they triangulate their surfaces from very large point datasets on the fly. The point locations are indexed, so that only the area being viewed is used to create a surface. Think of indexing as similar to tiling, or caching by area viewed, in ArcGlobe. Zoomed out at a small scale, a thinner selection of points is used, so speed is not impaired. Zoomed in to larger scales, more points are used to create more detail, but for a smaller area, so again, not much drawing needs to take place. Instead of housing multiple TINs at various levels of detail, it's faster (and a space-saver) to execute instructions that draw the terrain as needed.

Terrains were developed to make use of data sources such as LiDAR and SONAR, which are becoming more readily available to average users. LiDAR stands for Light Detection and Ranging, and is an instrument that measures distances to reflecting objects by emitting pulses of laser light and measuring the time that it takes to bounce back. An airborne LiDAR system uses GPS to measure the x, y, z coordinates of the moving LiDAR sensors in the air, and also uses an Inertial Measuring Unit to factor in the roll and pitch of the airplane. The LiDAR sensors emit between 5000 and 50,000 pulses per second. The datasets ultimately resulting from the returns can contain hundreds of thousands of points.

Mass points created from LiDAR returns

Not surprisingly, the best way to become familiar with the terrain data model is to build one. In this chapter, we'll convert LiDAR data to a multipoint feature class, use it build a terrain surface model, rasterize the terrain, and view it in ArcMap and ArcGlobe.

Exercise 8-1

Create a Terrain Dataset

In this exercise, you'll convert LiDAR data to a multipoint feature class, create a new geodatabase with a feature dataset, and build a terrain surface model.

Step 1. Examine the Data That You'll Be Using to Create the Terrain

Start ArcCatalog, and navigate to your 3DDATA\Chapter08\Data folder.

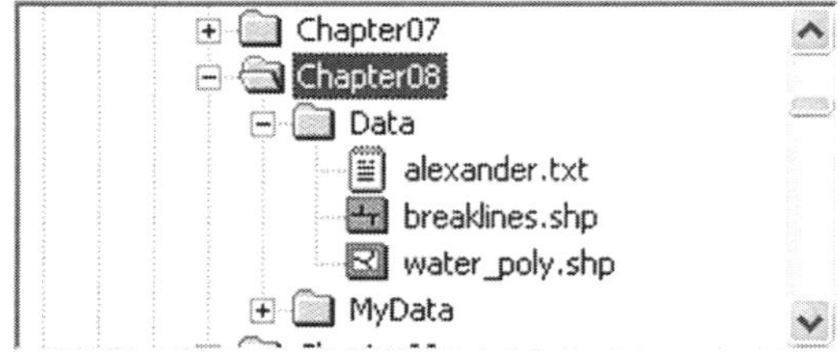

Breaklines.shp and water_poly.shp are shapefiles that control triangulation much as they would in a TIN. Take a look at the geography and the attribute tables of the two shapefiles, using the Preview tab. The breaklines are of geometry type ZM, and the water_poly has a field called "Height" that will be used to provide elevation values when we add it to the terrain.

In the Data folder click alexander.txt, and select the Preview tab.

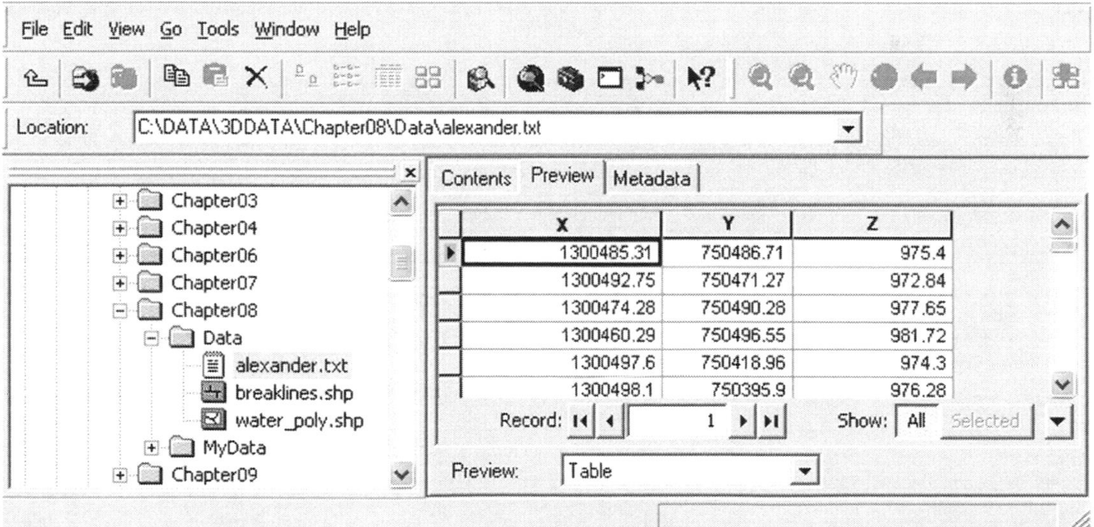

Alexander.txt is a comma-delimited text file containing tens of thousands of lines of x,y,z coordinates, collected from LiDAR sensors in Alexander County, North Carolina. The first line in the text file is X, Y, Z.

ArcCatalog correctly interprets this line as the column headings for the table, and parses the rest of the information into the proper fields.

ArcCatalog knows how to handle .txt files, and properly displays the contents of this file as long as it has a .txt extension. In order for the Terrain tools to recognize this LiDAR data, however, the file needs to have a .xyz extension. There is actually a copy of this file, called alexander.xyz, in the Chapter08\Data folder, but you won't see it unless you've customized ArcCatalog to show files of type "xyz." Even then, ArcCatalog wouldn't know how to show the contents of the file. Never fear, however; ArcToolbox will recognize alexander.xyz when you use it to create a multipoint feature class in the next few steps. Alexander.txt is only included here so that you can get a look at the LiDAR data in its raw form.

Step 2. Create a File Geodatabase and Empty Feature Dataset

A terrain resides inside a feature dataset, within a geodatabase, and is made of feature classes that "participate" in the terrain. All of the elements of the terrain need to be inside the feature dataset of the same geodatabase. The feature dataset provides a common coordinate system and spatial domain for terrain processing.

In ArcCatalog, make sure you're working in the 3DDATA\Chapter08 folder.

Right-click your MyData folder, point to New, and choose File Geodatabase.

When the new file geodatabase is created, rename it "Alex_terrain."

Right-click Alex_terrain, point to New, and choose Feature Dataset.

In the New Feature Dataset dialog, name the dataset "topo."

Click Next.

The next dialog asks you to choose a coordinate system.

Make sure "Projected Coordinate Systems" is highlighted.

Click the Import button.

Navigate to the Chapter08\Data folder, and highlight breaklines.shp. Click Add.

This imports the NAD 83 North Carolina State Plane projection for the feature dataset.

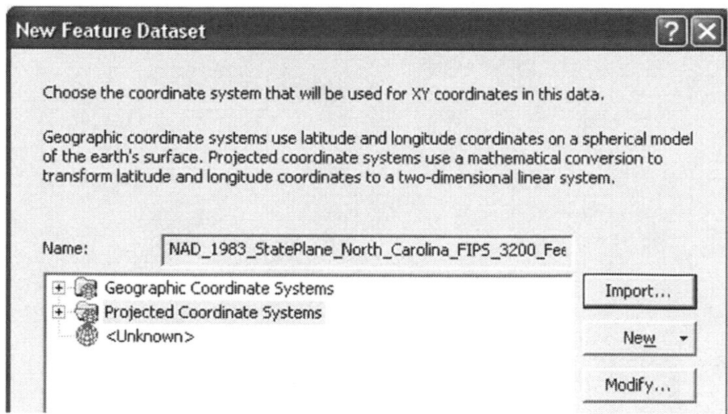

Click Next.

When the dialog asks you for a Z coordinate system, double-click "Vertical Coordinate Systems" to bring up the dropdown list.

In the dropdown list, double-click North America, and choose "NAVD 1988."

Click Next.

Accept the default parameters, and click Finish. ArcCatalog creates the empty feature dataset "topo" within the file geodatabase "Alex_terrain."

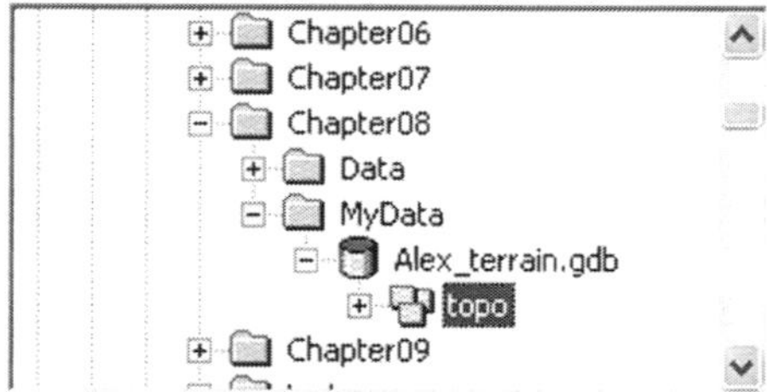

Step 3. Import Breaklines.shp and Water_poly.shp into the "Topo" Feature Dataset

In ArcCatalog, right-click "topo," point to Import, and choose "Feature class (multiple)."

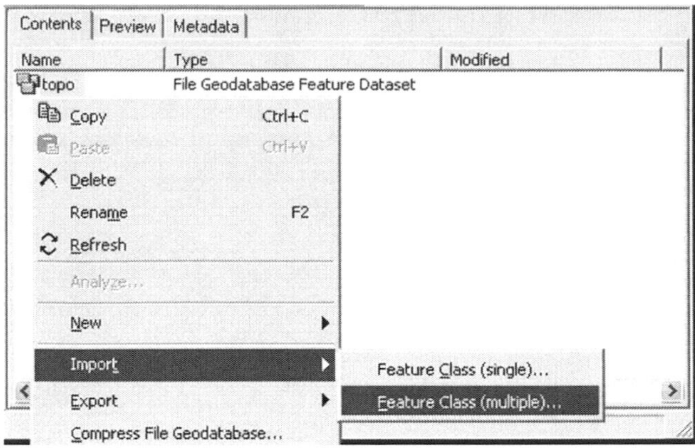

(Make sure that you are importing feature classes into the topo dataset, not just into the Alex_terrain geodatabase.)

The "Feature Class to Geodatabase (multiple)" dialog appears.

Click the Browse button next to Input Features. Navigate to your 3DDATA\Chapter08\Data folder. Hold down the Ctrl key, and highlight both breaklines.shp and water_poly.shp. Click Add.

The dialog should look like this, with your own initial pathnames, of course:

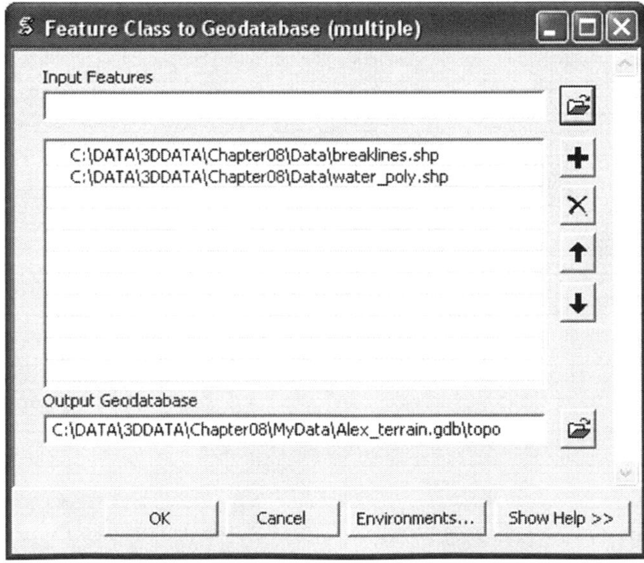

Click OK. ArcCatalog processes the request and adds breaklines and water_poly as feature classes to the topo dataset within Alex_terrain geodatabase, all inside your MyData folder.

When ArcToolbox signifies that it has finished processing the request, click Close on the dialog box.

Step 4. Load ArcToolbox

If ArcToolbox is not already visible, click the ArcToolbox button on the Standard toolbar in ArcCatalog.

Step 5. Convert the LiDAR Data to a Multipoint Feature Class

The alexander.txt file houses thousands of lines of text—one row for each LiDAR return point. It is much more efficient to group the points into clusters, creating a table with fewer rows in which each row is treated as a multipoint.

In ArcToolbox, double-click "3D Analyst Tools" to expand the toolbox.

Double-click "Conversion" to expand that toolbox; then expand the From File toolbox.

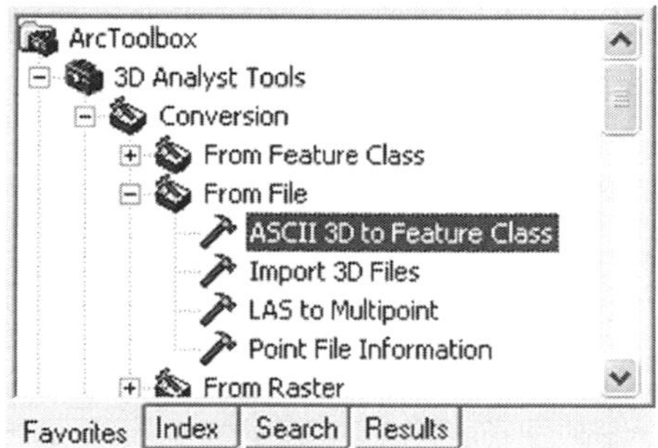

Finally, double-click "ASCII 3D to Feature Class" to open the tool.

In the ASCII 3D to Feature Class dialog, make sure that the "Browse for" option is set to "Files."

Click the Browse button in the Input panel, navigate to the 3DDATA\Chapter08\Data folder, and add "alexander.xyz."

Make sure the Input file format is set to "XYZ."

Click the Browse button next to the Output Feature Class option, and navigate through to 3DDATA\Chapter08\MyData\Alex_terrain.gdb\topo. Name the new feature class "mass_points."

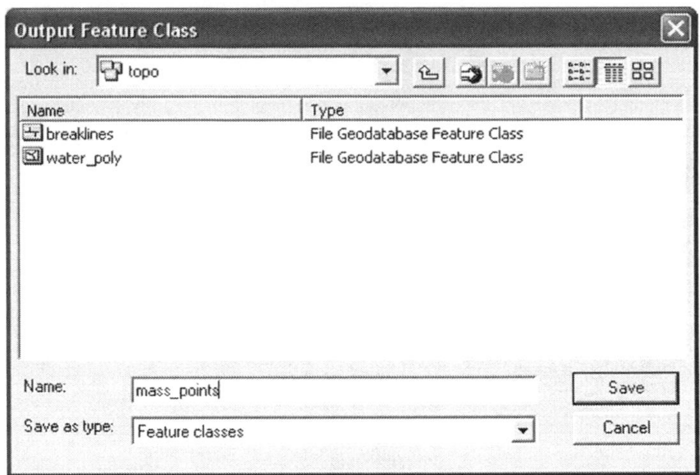

Click the Save button to return to the ASCII 3D to Feature Class dialog.

Make sure that the Output Feature Class type is set to MULTIPOINT.

In the Average Point Spacing box, type "16."

When you're finished, the dialog should look like this:

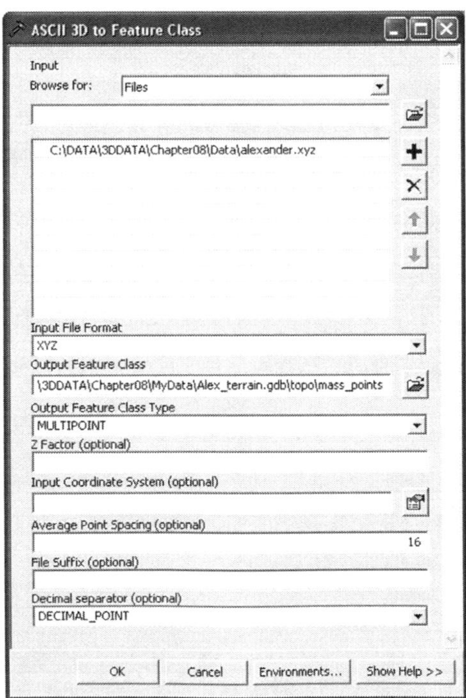

Click OK.

When ArcToolbox signifies that it has finished processing the request, click Close on the dialog box.

Step 6. Examine the New Multipoint Feature Class

Within the topo dataset, select "mass_points" and click the Preview tab. Set the Preview dropdown choice to "Geography" and zoom in to take a look at the mass points.

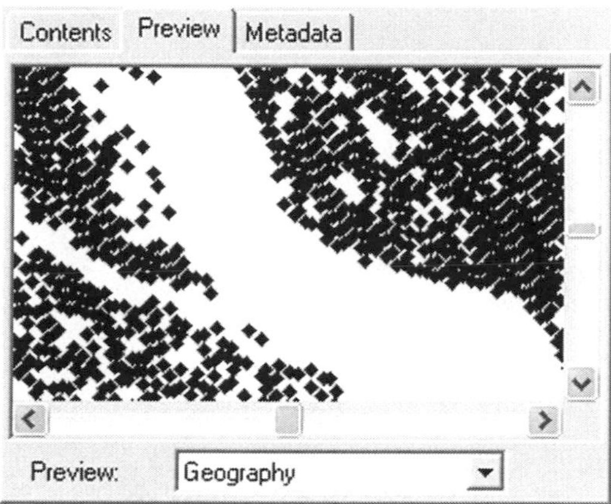

Change the Preview dropdown choice to "Table."

Instead of holding thousands of records in the table, the multipoint feature class only has 82 records. Each one represents up to 5000 grouped point features. (At the end of the table, you'll see that the grouping of points becomes less regimented, as the ASCII 3D to Multipoint tool gathers the last of the points together.) If you were to open mass points in ArcMap and select one record in the table, all of the points that that record represents would be selected in the view.

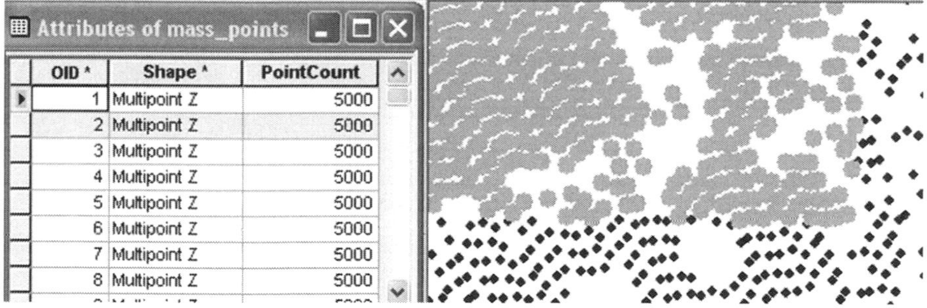

Attributes of mass_points

OID *	Shape *	PointCount
1	Multipoint Z	5000
2	Multipoint Z	5000
3	Multipoint Z	5000
4	Multipoint Z	5000
5	Multipoint Z	5000
6	Multipoint Z	5000
7	Multipoint Z	5000
8	Multipoint Z	5000

Step 7. Build the Terrain Dataset

In ArcCatalog, right-click "topo," point to New, and select Terrain.

Accept the new name "topo_Terrain," and click the Select All button to check on all three feature classes that will participate in the terrain.

Type "16" for the approximate point spacing.

Click Next.

On the next dialog, change the SF Type for water_poly to "hard replace."

Feature Class	Height Source	SFType
mass_points	Shape	mass points
breaklines	Shape	hard line
water_poly	Height	hard replace

Click Next.

Now, you'll create the pyramids that tell the terrain how much detail to display at a given scale.

In the dialog, click the radio button next to "Z Tolerance." This is generally the best choice for bare earth LiDAR, which is what we're using. The z-tolerance of each pyramid level is a measure of its vertical accuracy relative to the finest resolution of the data.

Click Next again.

The next dialog asks you to set the values for the pyramid levels.

Click "Calculate Pyramid Properties," and then click "Add" twice.

Alter the values so that the final table looks like this:

No.	Z Tolerance	Maximum Scale
1	0.5	5000
2	1.5	10000
3	2.5	25000

Click Next.

Your final result should reflect the dialog below:

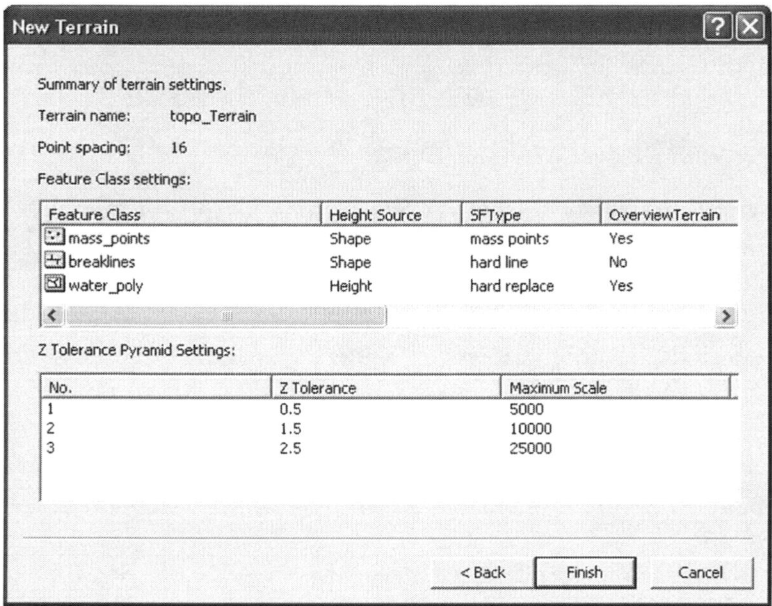

Click Finish.

When the terrain's settings have been created, the wizard asks you if you'd like to build the terrain.

Click Yes.

The building process takes a little while. When it's finished, topo_Terrain appears inside the topo dataset.

Step 8. Examine the New Terrain in ArcCatalog

Click topo_Terrain, and select the Preview tab.

Zoom in to take a closer look at your awesome new terrain. Notice that you have to zoom in past a scale of 1:5000 before you see the breaklines, because you chose "No" in the dialog that asked you if you wanted them to be part of the Overview Terrain. The breaklines are only visible at the terrain's full resolution scale. The water_poly feature class is visible at all three pyramid scales, however, because you chose to include it in the Overview Terrain.

When you're finished, close ArcCatalog. In the next exercise, we'll look at topo_Terrain in ArcMap, rasterize it, and then view it in ArcGlobe.

Exercise 8-2

Rasterize a Terrain Dataset and View it in ArcGlobe

Step 1. Start ArcMap and Add Data

Open ArcMap, and click the Add Data button. Navigate to your 3DDATA\Chapter08\MyData folder.

Double-click alex_terrain.gdb, double-click the topo dataset, and add topo_Terrain to the view.

At the full extent of the data, the Table of Contents should tell you that you're viewing the Overview Terrain. (If not, you can change that in the Terrain's layer properties, as explained in Step 2.)

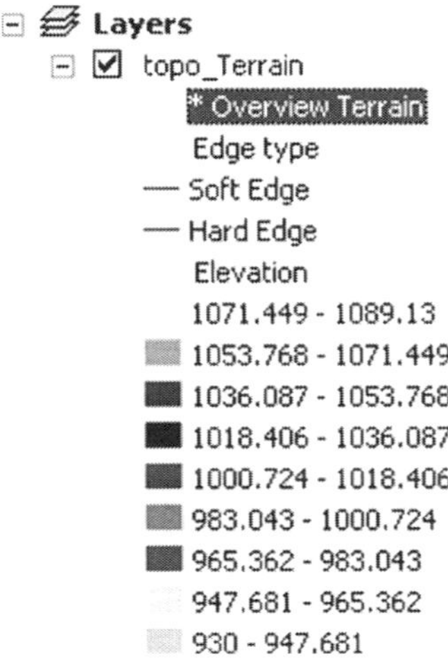

Zoom in to a scale of 1:10,000, and the Table of Contents shows that the Z-tolerance for this scale is 1.5.

Zoom in further to 1:5000, and the Z-tolerance changes to 0.5.

Zoom in any further than 1:5000, and the Z-tolerance becomes 0.00. These are the scales and Z-tolerance levels that you specified when you created pyramids for the terrain in the last exercise.

Step 2. Look at topo_Terrain's Display Properties

In the Table of Contents, double-click topo_Terrain. In the Layer Properties dialog, click the Display tab.

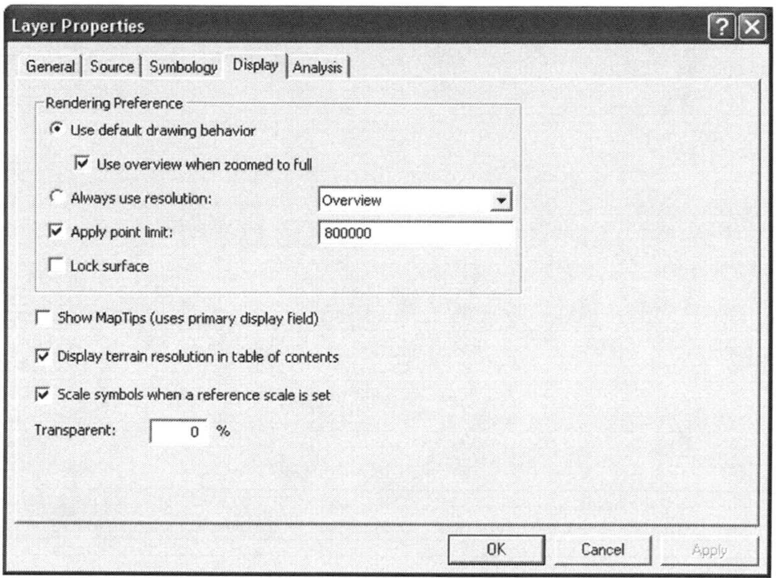

You're seeing the Z-tolerance settings in the Table of Contents because "Display terrain resolution in the table of contents" is checked on.

The Rendering Preference panel tells you that ArcMap is using the terrain's default drawing behavior—that is, according to the pyramid levels that you set earlier. You also have the choice of clicking the "Always use resolution" radio button, and setting the terrain so that it always displays at a resolution of 0.00, 0.5, 1.5, or 2.5 (the Overview level).

If "Use overview when zoomed to full," is checked, ArcMap will display the coarsest resolution at what it deems full extent, even though, in our case, we have specified that ArcMap use a Z-tolerance of 1.5 all the way out to a scale of 1:25,000.

Close the Layer Properties dialog.

Step 3. Convert the Terrain to an Elevation Raster

A terrain can be used as-is in ArcGlobe; it can provide elevation values (in which case it will not be visible) and it can be draped on the surface. However, ArcGlobe will rasterize the terrain on the fly, just as it does other vector features. In order to save this step, it's more efficient in many cases to convert the terrain to a new elevation raster. And if you want to view a terrain in ArcScene, it must be converted to a raster or a TIN.

If ArcToolbox is not loaded, click the ArcToolbox button on the Standard toolbar.

In ArcToolbox, expand the 3D Analyst Tools, then expand the Conversion toolbox, then From Terrain.

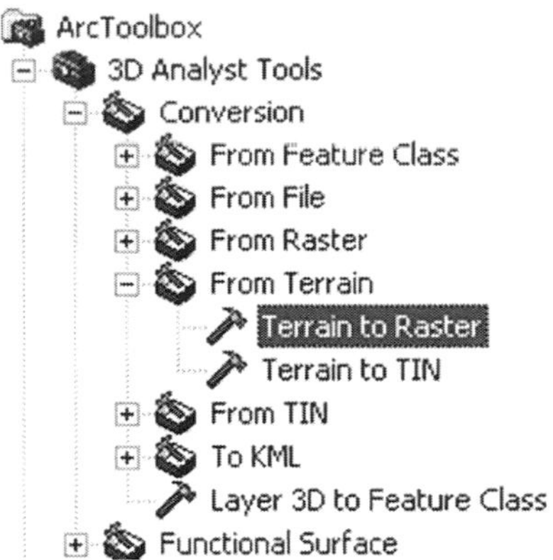

Open the Terrain to Raster tool by double-clicking it.

In the input dropdown list, choose topo_Terrain.

Call the output raster "terrain_rast" and save it in your 3DDATA\Chapter08\MyData folder. (Do not save it inside the Alex_terrain geodatabase.)

Set the output data type to Float, the method to Natural Neighbors, and the sampling distance to Cellsize 30. The pyramid level should be 0 (full resolution).

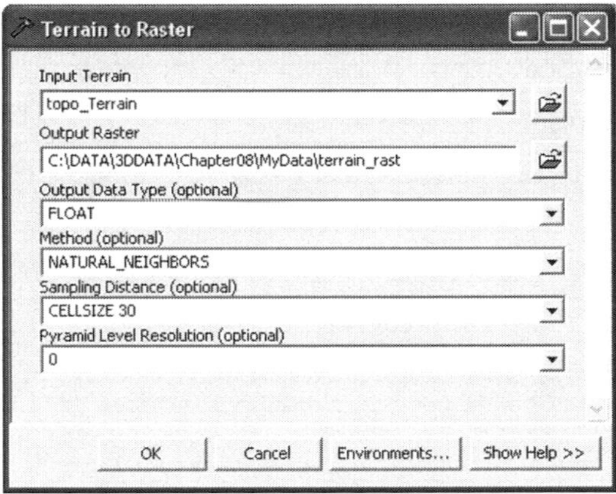

Click OK.

The process takes a few moments. Close the dialog when it has successfully completed.

Terrain_rast is added to ArcMap's Table of Contents. Turn off topo_Terrain so that you can see the new layer.

Step 4. Create a Hillshade of Terrain_rast

If the 3D Analyst toolbar is not loaded in ArcMap, click the View menu, point to Toolbars, and select 3D Analyst.

From the 3D Analyst menu, point to Surface Analysis, then Hillshade.

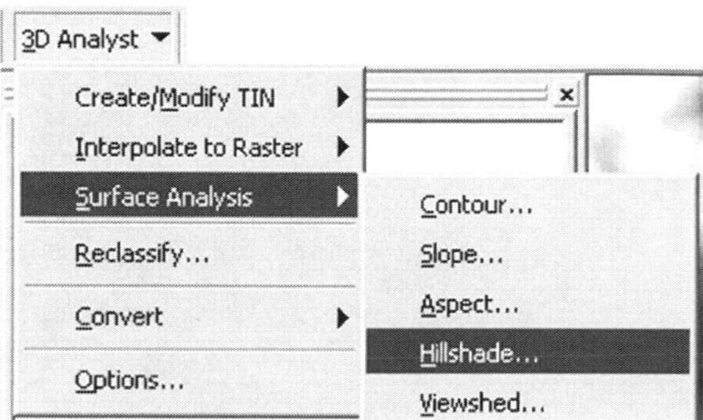

In the Hillshade dialog, the input raster should be terrain_rast and the output cell size should be 30. Call the output raster "terrain_hshd" and save it in your 3DDATA\Chapter08\MyData folder.

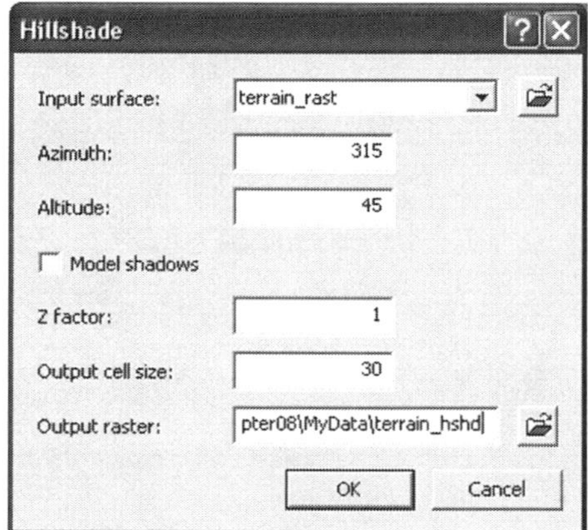

Click OK.

The new hillshade layer is added to the map.

In the Table of Contents, drag terrain_rast above terrain_hshd.

Double-click terrain_rast to bring up its layer properties.

On the Display tab, set the Transparency to 40%.

Under "Resample during display using," choose "Bilinear Interpolation."

On the Symbology tab, choose a color ramp that suits your taste.

Click OK to close the Layer Properties dialog, and take a look at the effect in ArcMap.

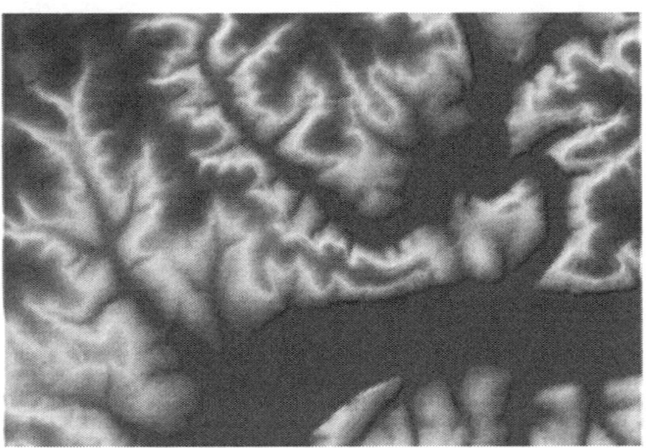

Step 5. Save Terrain_rast as a Layer File

You're going to add terrain_rast to ArcGlobe in the next few steps, and it would be nice not to have to set the color and transparency again.

In the Table of Contents, right-click terrain_rast, and click "Save as Layer File."

Save it in your 3DDATA\Chapter08\MyData folder, and name it "terrain_rast.lyr."

Investigate the new raster layers in the view. When you're finished, close ArcMap without saving changes.

Step 6. Add Terrain_rast as an Elevation Layer in ArcGlobe

You can certainly use your original terrain layer in ArcGlobe, either for elevation or as a draped layer. Instead, though, you'll use the rasters that you just created from the terrain. The terrain is about 10 megabytes, while the rasters are each less than half a megabyte.

Start ArcGlobe. At the bottom of the Table of Contents, make sure that the Type tab is selected.

Click the Add Data button, and navigate to your 3DDATA\Chapter08\MyData folder.

Add terrain_rast to the view (not terrain_rast.lyr).

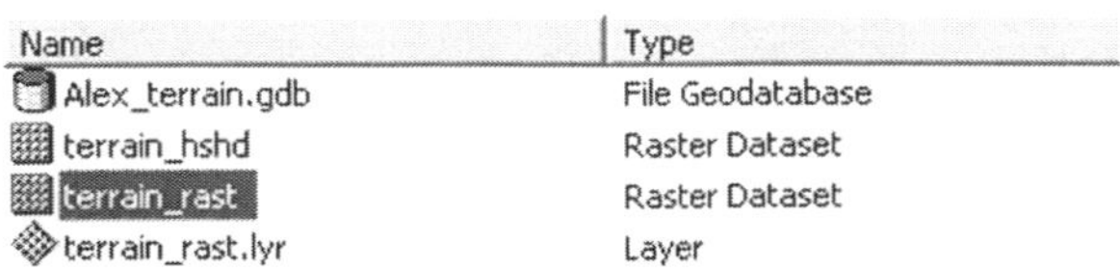

In the Add Data Wizard, click the radio button next to "Use this layer as an elevation source."

Click Finish.

When the Geographic Coordinate Systems Warning comes up, click the Transformations button.

In the next dialog, convert from GSC North American 1983 to WGS 1984 using method number 5 in the dropdown list.

Convert from:
GCS_North_American_1983
GCS_WGS_1984
OK
Cancel
Into:
GCS_WGS_1984
Using:
NAD_1983_To_WGS_1984_5
New...
Method: Coordinate Frame - dx=-0.991000 dy=1.907200 dz=0.512900
rx=-0.025790 ry=-0.009650 rz=-0.011660 s=0.000000

Click OK, and then click Close on the Transformations dialog.

Terrain_rast is added as an elevation layer to the globe. Make sure that it's listed in the Table of Contents at the top of the Elevation Layers category, right above "Elevation (30m)."

Step 7. Examine Terrain_rast

In the Table of Contents, right-click terrain_rast and choose Zoom to Layer.

The area looks familiar, but something's off. Terrain_rast's elevation values are creating a giant geographical loaf in the middle of a relatively flat area.

It looks about three times taller than it should. About 3.2810 times taller, as a matter of fact.

Step 8. Correct for the Feet to Meters Discrepancy

Right-click terrain_rast in the Table of Contents, and choose Properties.

On the Elevation tab, note that the layer is providing elevation to the globe's surface. In the Unit Conversion panel, click the dropdown arrow next to "Custom" and choose "feet to meters."

Unit Conversion
Factor to convert layer elevation values to meters: feet to meters 0.3048

Click OK.

Now terrain_rast fits in with the other elevation data on the globe. The elevation values in the original XYZ text file that you used to make the Terrain were in feet; thus, topo_Terrain's elevations were in feet, as are those of terrain_rast. ArcGlobe's default elevation layers (at version 9.3) are in meters, so it gives you the option of correcting the layer's elevation values by converting feet to meters.

Step 9. Add Terrain_rast.lyr and Terrain_hshd as Draped Layers

Terrain_rast is providing elevation values, so it's invisible in ArcGlobe. If you want to see it, you have to add it again as a draped or floating layer. You'll add the .lyr file that you made in ArcMap, since it's already symbolized to your specifications.

In the Table of Contents, right-click "Globe Layers" and choose Add Data, then Add Draped Data.

Navigate to your 3DDATA\Chapter08\MyData folder, and add both terrain_rast.lyr and terrain_hshd.

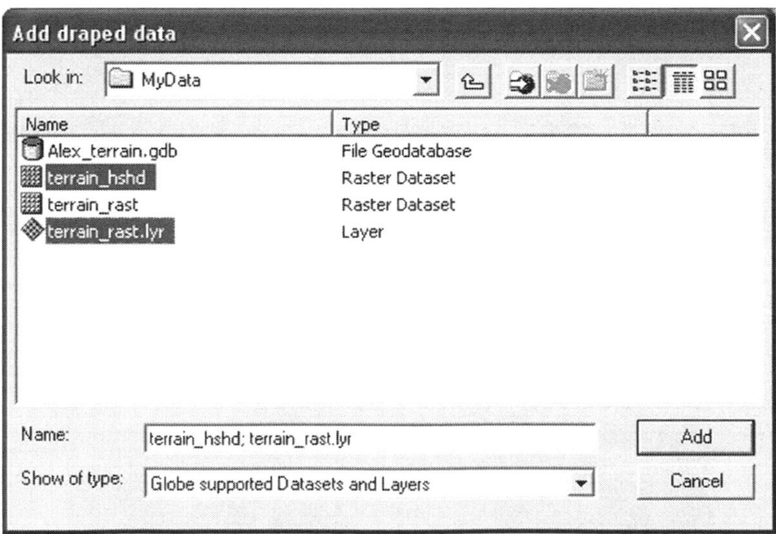

Use the same Transformation settings in the Coordinate Systems warning dialog as you did for terrain_rast in Step 6.

The two new layers are added to the view, but the Imagery layer may be covering them up.

Drag terrain_rast.lyr so that it is above the Imagery layer in the Table of Contents.

Zoom in and take a look at the effect. Since terrain_rast.lyr has a transparency set, you can see aspects of the Imagery layer underneath.

Now drag terrain_hshd in the Table of Contents so that it resides above the Imagery layer but below terrain_rast.lyr. Note the change in the effect.

Experiment with different settings for your new layers. Try adding topo_Terrain from the Alex_terrain geodatabase. When you're through, close ArcGlobe, saving the changes to a new ArcGlobe document if you like.

CHAPTER 9

3D Features and More Surface Analysis Techniques

Now that you understand the basics of raster, TIN, and terrain data structures, and you're familiar with the workings of ArcScene and ArcGlobe, it's time to reward yourself with the simple activity of drawing 3D shapefiles in ArcMap. You'll also learn how to calculate 3D area and volume statistics, draw a line of sight along a surface, and create a cross-section profile of that line.

3D Feature Classes: Shape-Z and Multipatch

We discussed 3D vector features in Chapter 1, so you may remember that 3D points, lines, and polygons are just like their 2D counterparts, except for the fact that they store elevation z values. A point has one z value, while lines and polygons store a z value at each vertex.

There are two basic ways to create 3D features. You can either digitize them in ArcMap, using a TIN or an elevation raster as the source for z values, or you can convert existing 2D features to 3D. When you convert features to 3D, you can get the z values from a TIN or raster surface, from a field containing elevation values in the feature attribute table, or you can set a single value that assigns a constant elevation to all the vertices in the feature layer.

Multipatch features can also be created in a variety of ways. CAD data can be imported and viewed as is; 2D polygon features can be interpolated to a multipatch feature class, using a TIN surface for the Z values; and 3D models in other formats such as 3D Studio, VRML, OpenFlight, and Sketchup can be imported to multipatch features within a geodatabase. You can also symbolize point features with 3D markers, save that as a layer (.lyr) file, then import that layer file into a multipatch feature class in a geodatabase.

3D Analysis Techniques

The steepest path tool models the route that a liquid such as water or lava will take on a slope. You can also use it to test the accuracy of your surface model. The more accurate the model, the more closely the modeled path of a water source will line up with the course that the actual stream bed takes.

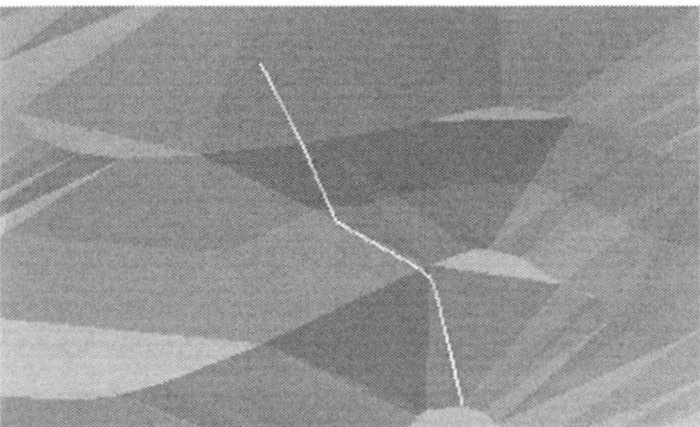

The white line represents the path water should take from its origin at the top of the slope

Profile graphs show a cross-section of a surface along a line that you draw, charting the distance of the line on horizontal and vertical axes. This is a function of 3D Analyst, but it can only be done in ArcMap.

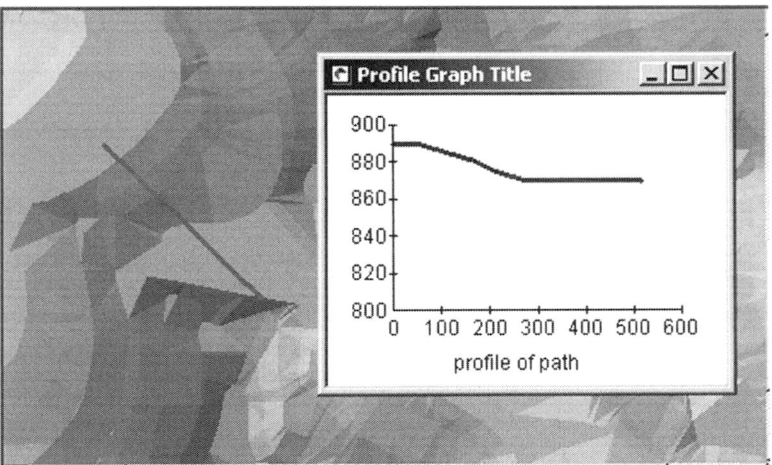

The graph shows the height and distance, in feet, of the line drawn along the surface of the TIN

A line of sight is related to the targets and observers that you studied in Chapter 3, and the viewsheds you studied in Chapter 6. In this case, you choose an observer and a target, and 3D Analyst draws a line directly between them, indicating which parts of the surface along the line are visible to the observer and which are not. Like the profile graph, the line-of-sight tool only works in ArcMap.

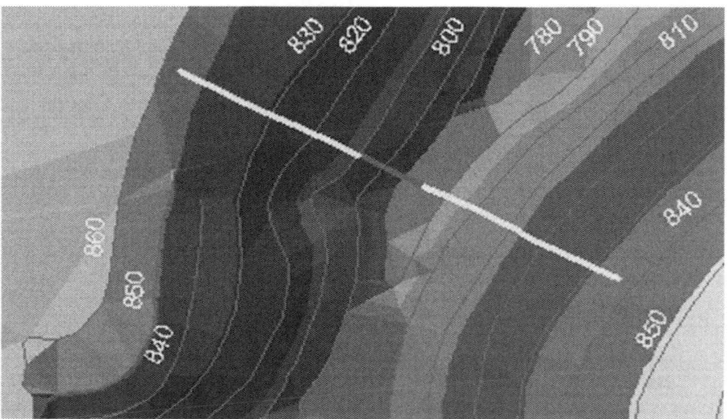

The line drawn across the gully between two points shows not only whether the target is visible but also which parts of the surface along the line are hidden from view. In the graphic, the light sections of the line (green) show where the surface is visible. The dark section (red) shows where the view of the surface is obstructed

Area and volume statistics calculate a surface's two-dimensional area, the surface's actual area (lumps and all) and the volume of material between the top or bottom of the surface and a reference plane that you specify. These numbers are useful for applications such as cut and fill, damming, and lake construction.

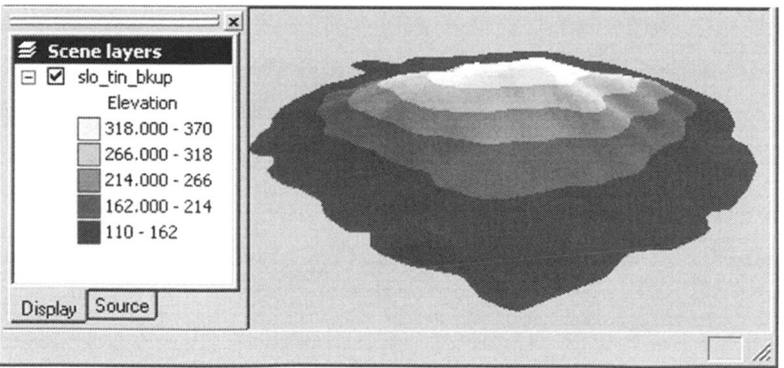

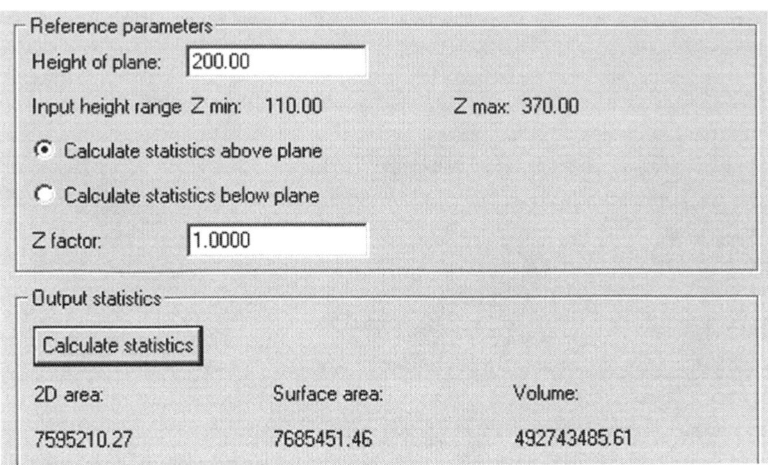

These statistics are calculated for the TIN above from an elevation plane of 200 feet to the top of the hill. Notice that the surface area is greater than the 2D area. Unless a surface is flat, the actual surface area will always be greater than the 2D area

In the following exercises, you'll convert 2D features to 3D, create a 3D shapefile in ArcCatalog, digitize in ArcMap, and create a multipatch feature class. You'll also find area and volume, draw a line of sight, and create a profile graph.

Exercise 9-1

Convert 2D Features to 3D, and Digitize 3D Features in ArcMap

Generally, you make 3D vector features for the sake of convenience—so that you don't have to set their base heights to another layer when you add them to ArcScene. In this exercise, you'll convert existing 2D line and polygon features to 3D using a TIN of a rural area near Brushy Creek in Carlisle, Kentucky. You'll also create a new 3D shapefile using ArcCatalog and ArcMap.

Step 1. Start ArcScene and Add Data

Open ArcScene, and click the Add Data button. Navigate to your 3DDATA\Chapter09\Data folder and add brushy_creek_tin, buildings.shp, and roads.shp.

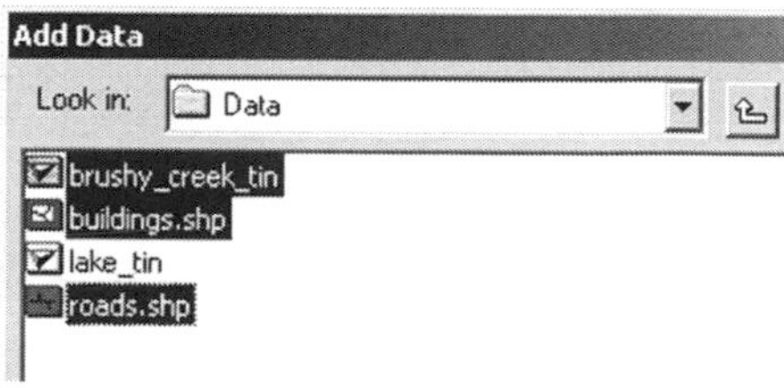

Step 2. Examine the Data

Double-click brushy_creek_tin in the Table of Contents to bring up its Layer Properties. Click the Symbology tab.

In the Show box, remove Faces. Click Add, and choose "Face Elevation with graduated color ramp."

Click the other Add button, then Dismiss. Click OK to close the Layer Properties dialog.

Navigate around the TIN and the two shapefiles. The roads and buildings are 2D features and don't have their base heights set to the TIN, so they float below it.

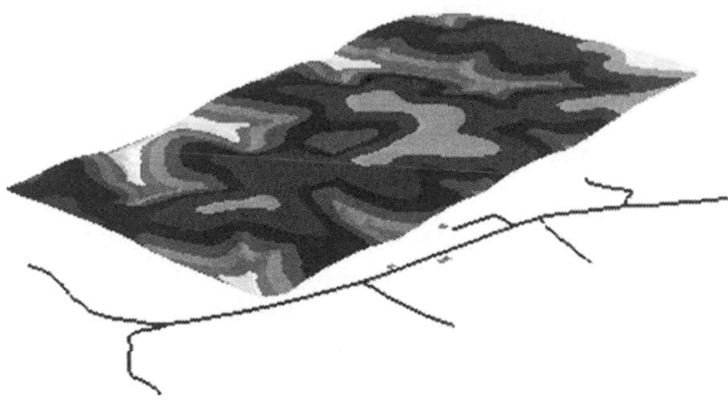

Zoom in and take a look at the depression in the middle of the TIN, outlined here in white. Later on, you'll be digitizing a lake polygon to fill this basin.

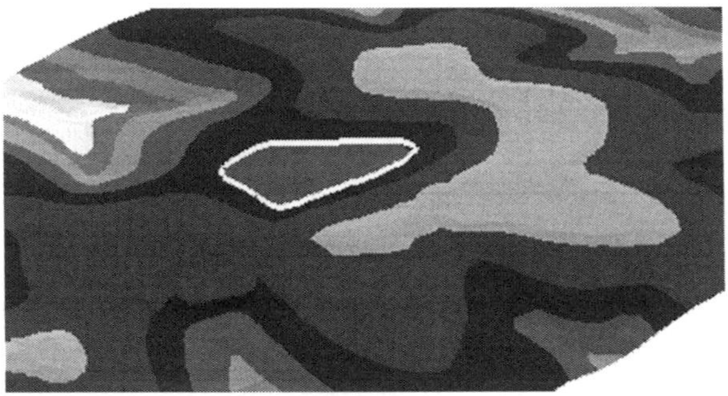

Step 3. Look at the Shapefile Attribute Tables

In the Table of Contents, right-click roads and choose Open Attribute Table.

Also open the attribute table for the buildings layer.

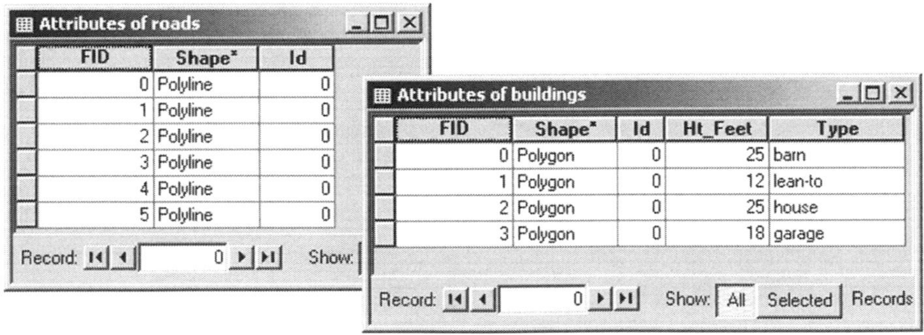

Attributes of roads

FID	Shape*	Id
0	Polyline	0
1	Polyline	0
2	Polyline	0
3	Polyline	0
4	Polyline	0
5	Polyline	0

Record: 0 Show:

Attributes of buildings

FID	Shape*	Id	Ht_Feet	Type
0	Polygon	0	25	barn
1	Polygon	0	12	lean-to
2	Polygon	0	25	house
3	Polygon	0	18	garage

Record: 0 Show: All Selected Records

You can tell that the features are 2D because they don't have a Z in the Shape field. Notice the Ht_feet field in the polygon table. You could use these values to convert the buildings to 3D, but they'd still be mighty short compared to the elevation above sea level of the TIN. So you'll use the TIN to convert the buildings to 3D, and later you'll use the Ht_Feet values to extrude the buildings from the TIN's surface.

Close the attribute tables.

Step 4. Convert the Building and Road Features to 3D

From the 3D Analyst menu, choose Convert, and then Features to 3D.

Examine the dialog for a minute. It looks a lot like the Base Heights tab on a Layer Properties dialog. You're familiar with setting base heights for vector features; you know that depending on what source you have, you can set them by typing in a constant value, by using the calculator button to assign values from a field in the feature layer's attribute table, by assigning the elevation values from a TIN or raster surface, or by using existing z values in the feature layer's Shape field. Here, you have the same choices—except for the last one, of course, because the purpose of converting features to 3D is to create a new layer that has z values in its Shape field.

In the dialog box, the Input features should be buildings. Set the source of heights to "Raster or TIN surface," and make sure that it's using brushy_creek_tin.

Click the Browse button next to the Output features box. Navigate to your Chapter09\MyData folder, and call your new shapefile 3d_buildings.shp.

Click the Save button. When you're done, the dialog should look like this:

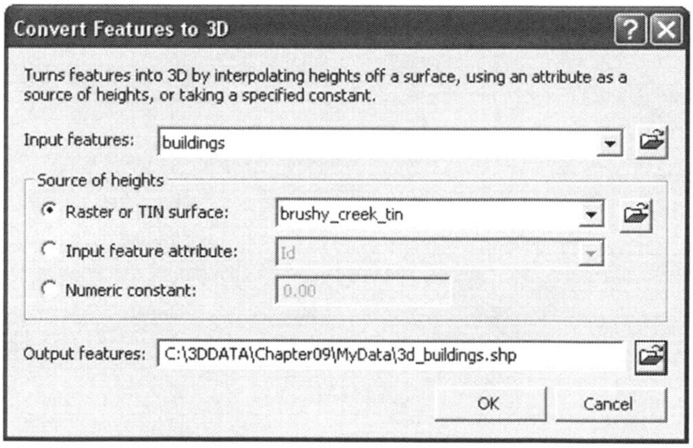

Click OK.

The new shapefile is added to the scene.

Repeat Step 4 for the roads shapefile, but name it 3d_roads.shp.

The new roads shapefile is also added to the scene. Now the features lie on top of the TIN.

Step 5. Examine the New Layers

In the Table of Contents, remove the original roads and buildings shapefiles.

Click the line symbol under 3d_roads. In the Symbol Selector, change the line width to 2 and choose a light color that will stand out against the TIN.

Change the color of the building features too, if you like.

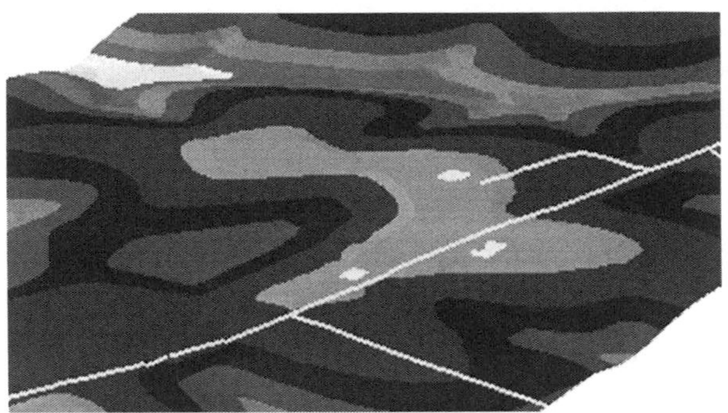

Open the attribute tables of roads and buildings. Now the Shape fields contain "PolylineZM" and "PolygonZM." The Z stands for elevation in this case, although a z value can be anything other than x, y coordinates. The M stands for measured values, which you get every time you convert 2D features to 3D, whether you specifically ask for them or not.

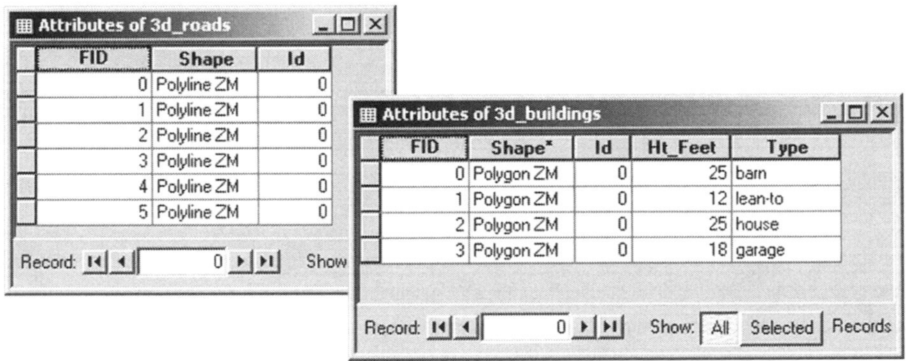

Attributes of 3d_roads

FID	Shape	Id
0	Polyline ZM	0
1	Polyline ZM	0
2	Polyline ZM	0
3	Polyline ZM	0
4	Polyline ZM	0
5	Polyline ZM	0

Record: 0 Show

Attributes of 3d_buildings

FID	Shape*	Id	Ht_Feet	Type
0	Polygon ZM	0	25	barn
1	Polygon ZM	0	12	lean-to
2	Polygon ZM	0	25	house
3	Polygon ZM	0	18	garage

Record: 0 Show: All Selected Records

Close the attribute tables.

Step 6. Extrude the 3D Building Polygons

Zoom into the buildings on the TIN surface.

In the Table of Contents, double-click 3d_buildings to bring up its Layer Properties.

On the Extrusion tab, click the Calculator button to bring up the Expression Builder.

In the Fields box, click Ht_Feet. It should appear in the Expression box.

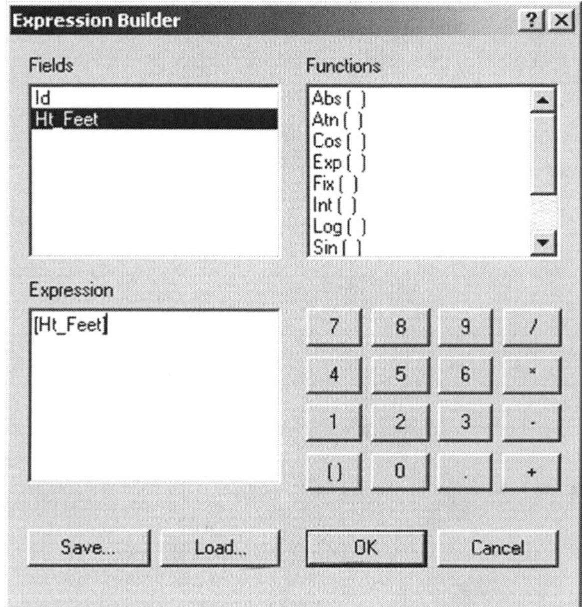

Click OK.

Back at the Extrusion tab, make sure that you are adding the extrusion to each feature's minimum height.

Extrude features in layer. Extrusion turns points into vertical lines, lines into walls, and polgons into blocks.

Extrusion value or expression:

[Ht_Feet]

Apply extrusion by:

adding it to each feature's minimum height

Click OK.

Now the 3D buildings have the correct base heights, set to their z values, and they're extruded to their building heights—25 feet for the barn and the house, 18 for the garage, and 12 for the lean-to.

Step 7. Save the Scene

From the File menu, click Save. Navigate to your 3DDATA\Chapter09\MyData folder, and give the scene a name of your choice. You're going to be working in ArcCatalog and ArcMap, so you can either close ArcScene or leave it open and minimize it, depending on your system resources.

Step 8. Create a 3D Shapefile of a Lake in ArcCatalog

Open ArcCatalog. Navigate to your 3DDATA\Chapter09\MyData folder.

Right-click the MyData folder. Select New, then Shapefile.

In the Create New Shapefile dialog, change the Name to 3d_lake, and for the Feature Type select Polygon.

Click the Edit button.

In the Coordinate System dialog, click the Import button.

Navigate to 3DDATA\Chapter09\MyData. Select either 3d_roads or 3d_buildings, and click the Add button. (If you don't have these shapefiles, use one of their backups from your Chapter09\Data folder.)

The Coordinate System panel is updated, showing that you've imported the NAD_83 Kentucky State Plane reference system to your new shapefile.

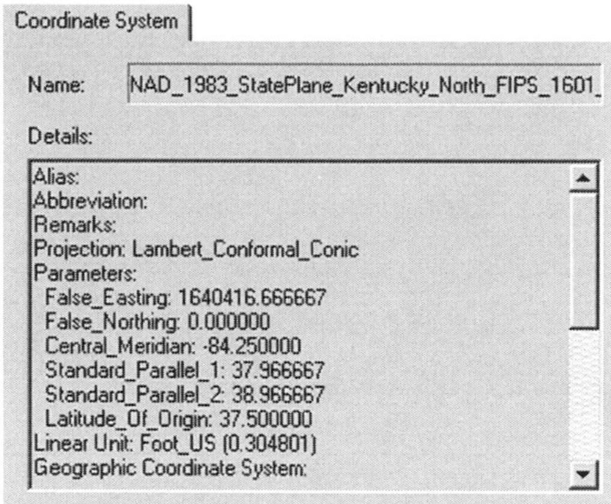

Click OK.

Back in the Create New Shapefile dialog, check the box to give the coordinates z values. When you're done, the dialog should look like this:

Create New Shapefile
Name: 3d_lake
Feature Type: Polygon
Spatial Reference
Description:
Projected Coordinate System:
Name: NAD_1983_StatePlane_Kentucky_North_FIPS_16
Geographic Coordinate System:
Name: GCS_North_American_1983
Show Details
Edit...
Coordinates will contain M values. Used to store route data.
Coordinates will contain Z values. Used to store 3D data.
OK
Cancel

Click OK.

A new, empty shapefile called 3d_lake is added to your MyData folder.

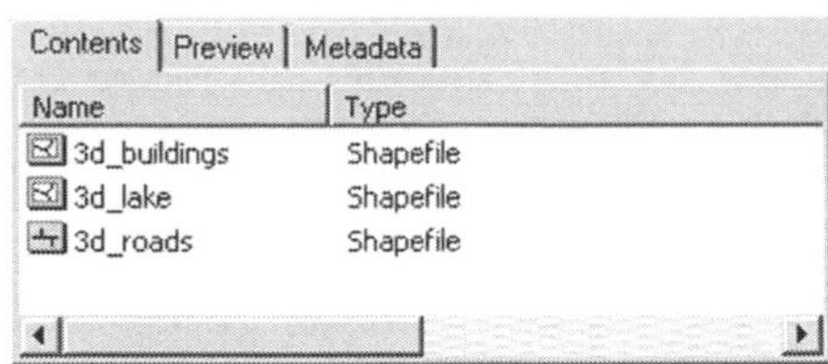

Close ArcCatalog.

Step 9. Open ArcMap and Add Toolbars

Open ArcMap.

You may already have the Editor toolbar, the 3D Analyst extension, and the 3D Analyst toolbar loaded in ArcMap, but if not:

From the Tools menu, choose Extensions. Make sure that the box next to 3D Analyst is checked.

From the View menu, choose Toolbars. Make sure that the boxes next to Editor and 3D Analyst are checked.

Step 10. Add Data

Click the Add Data button, and navigate to your 3DDATA\Chapter09\MyData folder.

Add the 3d_lake shapefile.

The new shapefile is added, but you don't see anything because it contains no features.

Click the Add Data button again. This time navigate up a directory to your 3DDATA\Chapter09\Data folder, and add lake_tin.

Zoom to the extent of lake_tin.

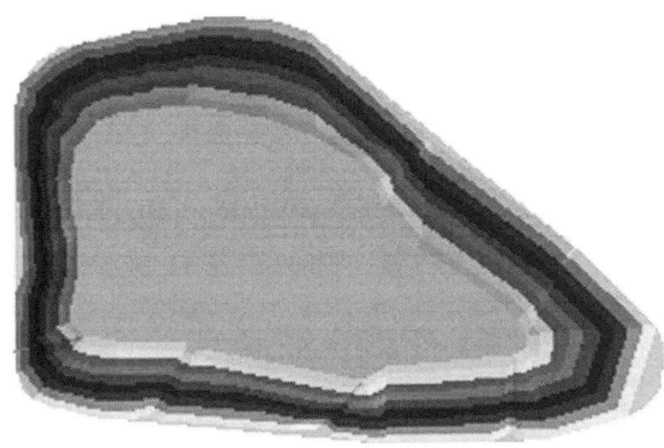

Lake_tin is a TIN of the lake that you looked at on the brushy_creek TIN earlier, in ArcScene. You'll use the outer edge to digitize the 3d_lake polygon.

Step 11. Interpolate a 3D Polygon

From the Editor menu, click Start Editing. The Target in the Editor toolbar should be 3d_lake.

On the 3D Analyst toolbar, click the Interpolate Polygon button. 

Put your cursor in the view.

The Interpolate tools are interesting. Unlike the regular Sketch edit tool, they rely on a 3D surface to get the z values for each vertex that you digitize. So when you digitize the polygon, it has to lie on or inside the boundary of the TIN. In fact, if during your sketch you wander outside the TIN, the Interpolate Polygon tool will automatically snap the vertex to the TIN's boundary. This creates an easy way to quickly make a perfect outline of the lake: if you digitize entirely outside the TIN, then no matter how quickly or sloppily you draw, the polygon will snap to the TIN, like this:

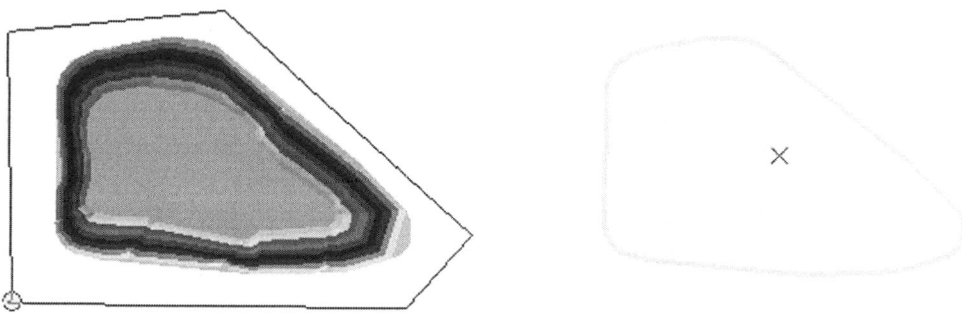

Click once to start drawing. When you've drawn the polygon, double-click the last vertex to finish the sketch. (If you make a mistake, double-click to finish the polygon, then delete it and start over.)

From the Editor menu, choose Save Edits, and then Stop Editing.

Step 12. Close ArcMap

Give the document a name of your choice and save it in your 3DDATA\Chapter09\MyData folder. Then close ArcMap.

Step 13. Open the ArcScene Project, and Add the 3D Lake

Maximize or reopen the ArcScene project that you saved in Step 7. It should contain brushy_creek_tin, 3d_roads.shp, and 3d_buildings.shp.

Click the Add Data button, and from your 3DDATA\Chapter09\MyData folder, add 3d_lake.shp.

The lake appears with its base heights set to the perimeter of the basin in the middle of the Brushy Creek TIN. The reason it isn't indented like the TIN you digitized it from is that polygons only carry z values at their vertices, and the vertices are only around the edge. 3D polygons work best for smooth surfaces like lakes and parking lots.

Step 14. Symbolize the Lake and the TIN

In the Table of Contents, click the symbol under 3d_lake to bring up the Symbol Selector.

Scroll down to "Water Body" in the main panel, and click OK.

Now double-click brushy_creek_tin in the Table of Contents to bring up its Layer Properties.

On the Symbology tab, choose a less drastic color ramp, such as Green Light to Dark. (You remember that you can right-click the color ramp and uncheck Graphic View to get the list of color ramp names.)

Click OK.

Step 15. Check Out the New Scene

Now the lake, the buildings, and the roads stand out nicely from the TIN surface.

Step 16. Close ArcScene

Feel free to experiment with symbology, base heights, and extrusion if you like. When you're done, save the document and close ArcScene.

Exercise 9-2

Draw a Line of Sight and a Cross-section Profile Graph

In this exercise, you'll create lines of sight, draw 3D graphics, and make a profile graph to illustrate the visibility results from the Line of Sight tool.

Step 1. Open ArcMap and Add Data

Open ArcMap, click the Add Data button, navigate to your 3DDATA\Chapter09\Data folder, and add brushy_creek_tin and lakeview.shp.

Click the Add Data button again, and this time add 3d_buildings from your 3DDATA\Chapter09\MyData folder. (If you don't have this file, use its backup from the Chapter09\Data folder.)

Step 2. Symbolize Lakeview.shp

Lakeview is a point shapefile that will serve as a visual marker when you draw a line of sight later in the exercise.

Click the lakeview point symbol in the Table of Contents to bring up the Symbol Selector. Increase the size to 8 or 9, and choose a bright yellow shade.

Step 3. Set Layer Properties for the 3D Buildings

In the Table of Contents, double-click 3d_buildings to bring up its Layer Properties.

Click the Labels tab.

In the Label Field select Type from the dropdown list.

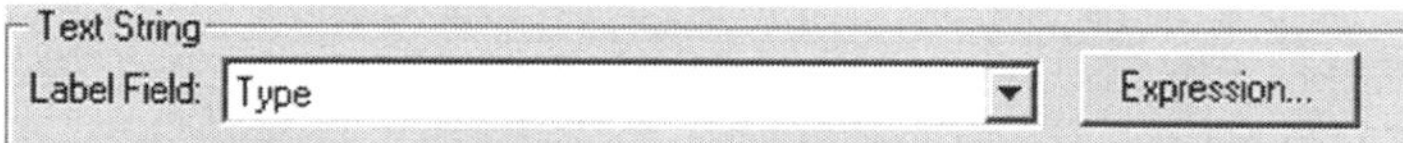

Now click the Symbology tab. In the Show box, click Categories. Unique Values is automatically highlighted.

In the Value Field dropdown list, choose Ht_Feet.

Under the main Symbol panel, click Add All Values.

Choose a color scheme that you like, preferably something bright that will stand out against the TIN.

In the Symbol panel, uncheck the box next to <all other values>.

When you're done, the Symbology tab should look something like this:

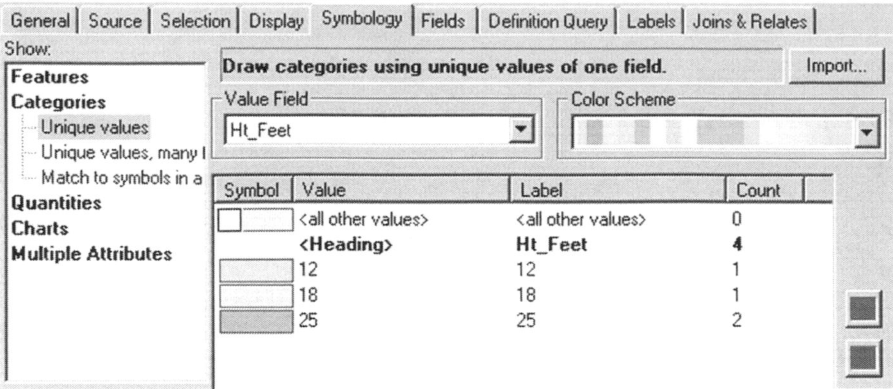

Click OK.

In the Table of Contents, right-click 3d_buildings and choose Label Features.

Zoom to the extent of the 3d_buildings layer.

Now the buildings are labeled, and classified by their height in the Table of Contents. This will come in handy when you set the target and observer elevations.

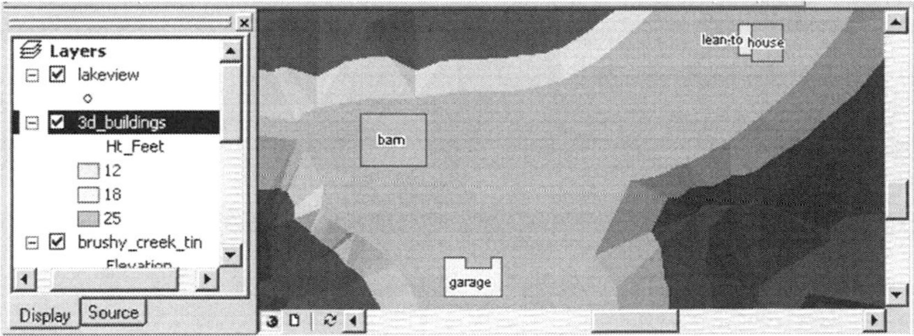

Step 4. Draw a Line of Sight

On the 3D Analyst toolbar, click the Line of Sight button. 

The Line of Sight dialog pops up.

Change the observer's offset to 25. Since the z units are feet, this puts your observer at the height of the barn.

Change the target offset to 18 feet, which is the height of the garage.

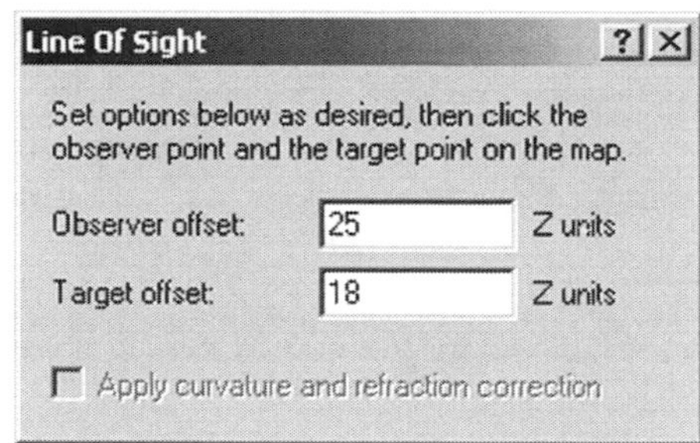

Now place your cursor on the barn, and click once. This sets the observer.

Place your cursor on the garage, and click once. This sets the target and draws the line of sight.

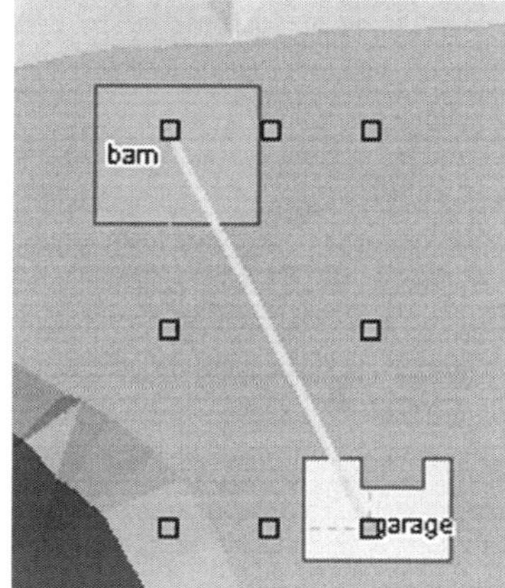

Without moving the cursor, look down in the left-hand corner of the Status Bar. It reports that the target is visible. (If you move the cursor out of the view, the Status Bar is cleared and you have to recreate the line-of-sight graphic to get another visibility report.) Not surprisingly, the line of sight is green, meaning that you can see the ground along the full length of the line.

Click the Delete key on the keyboard to remove the line-of-sight graphic.

(If you've clicked anywhere else in the scene, or moved your cursor out of the scene, the Delete key won't delete the graphic. In that case, select the Pointer tool (also called the "Select Elements" tool)

and click on the graphic. When you see little handles around the graphic, you know it's selected and can therefore be deleted.)

Step 5. Draw Another Line of Sight

Click the Line of Sight button again, to bring up the dialog.

This time, leave the observer's offset at 25, but change the target's offset to 100.

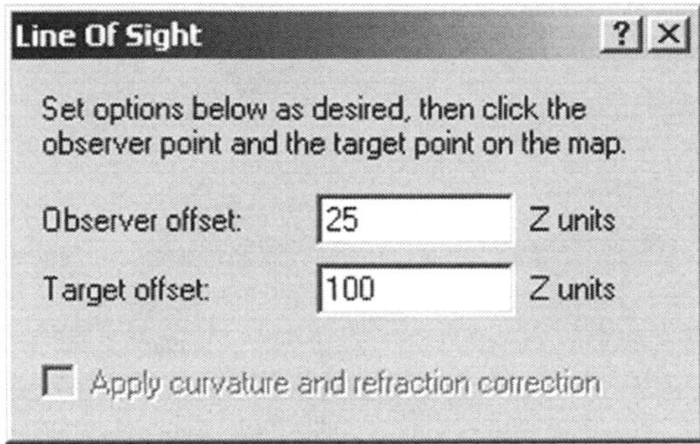

Click the barn for the observer, and click the garage for the target.

Again, the entire line is green, and the Status Bar reports that the target is visible. Although you might think that you couldn't see a target 75 feet on a rooftop above you, the Line of Sight tool isn't measuring whether you can see the top of the building or not. Since the target and the observer are on the same plane, 3D Analyst has no indication of any variation in the TIN surface that could obstruct the observer's view of the target. As far as it is concerned, on a flat plane you can see a point in the air 75 feet above you. Therefore, it's always good to do a few tests with the Line of Sight tool, so you know exactly what's being measured.

Click the Delete key to remove the line of sight graphic.

Step 6. Zoom Out to the Lakeview Shapefile, and Use the Identify Tool to Check Points of Elevation

Use the Pan and Zoom Out tools so that you can see the buildings and the lakeview point in the map.

Lakeview represents a bench located on the side of the pond opposite the farm buildings. Before you use the Line of Sight tool again, you should check the elevation of your target and observer.

Click the Identify tool. 

Click on the TIN surface, right next to the lakeview point. The Identify Results dialog pops up and tells you that the surface there is at an elevation of 850 feet.

Now click on the TIN surface right next to the barn. The barn sits at an elevation of 890 feet.

Close the Identify Results dialog.

Step 7. Draw Some More Lines of Sight

Click the Line of Sight button again. This time, set the observer offset to 4 feet (you're sitting on the lakeview bench) and set the target offset to 25 feet.

Click on the lakeview point, and then click on the barn.

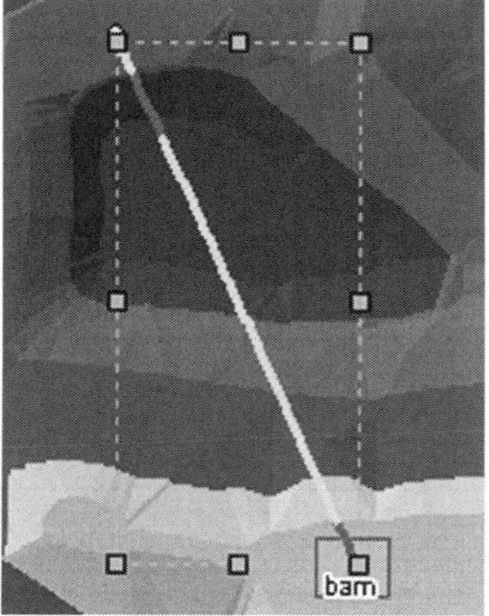

The Status Bar tells you that the target is visible. The line indicates that the down-slope leading into the pond is not visible, and that the ground in front of the barn isn't visible, either. But the offset of 25 feet makes the barn visible.

Click the Delete key to remove the graphic.

Click the Line of Sight button one more time. Leave the observer offset at 4 feet, but change the target offset to 2 feet.

Click on the lakeview point, and then click on the barn.

This time, the red and green portions of the line are the same, but the Status Bar reports that the target is not visible. The slope leading up to the target cuts off your vision of the bottom 2 feet of the barn.

Step 8. Experiment with the Line of Sight Tool

By changing the length of the line of sight and the target offset, you can get a feel for the cutoff point of visibility. For example, if you change the target offset to 10 feet, and draw the same line of sight from the lakeview point to the barn, the target is reported as visible. (You can see 10 feet up the side of the barn.) But if you extend the line of sight well past the barn with a target offset of 10 feet, the target is not visible. That's because the angle of the slope leading up to the barn is too steep—a point that will be illustrated more clearly in the next part of the exercise.

When you finish working with the Line of Sight tool, close the Line of Sight dialog. If you have a bunch of graphics left over in the scene, click the Edit menu and choose Select all Elements. Then click the Delete key to remove the graphics from the map.

Step 9. Interpolate a 3D Graphic Line

In the next step, you'll create a profile graph of the line that you've been drawing between the lakeview point and the barn. Before you do that, however, you have to interpolate a 3D graphic line; the Line of Sight tool only creates 2D graphics.

On the 3D Analyst toolbar, click the Interpolate Line tool. 

Click once on the lakeview point, then double-click on the barn.

A 3D graphic line is created.

Step 10. Make a Profile Graph

A profile graph is a two-dimensional cross-section of a 3D surface along a line that you select. The Profile Graph feature works with 3D lines or 3D graphics and is available only in ArcMap.

Make sure that the 3D graphic is still selected.

On the 3D Analyst toolbar, click the Create Profile Graph button. 

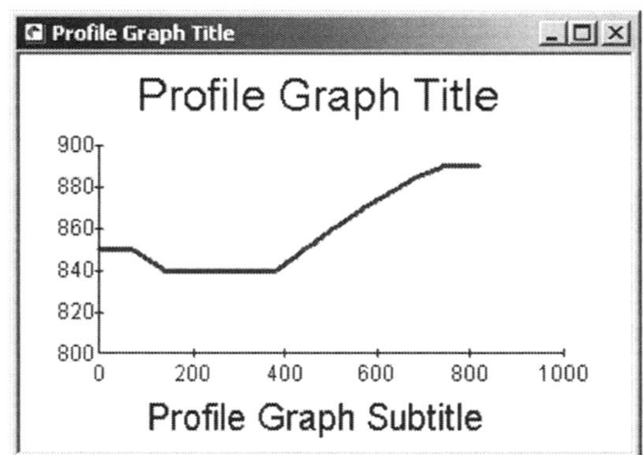

3D Analyst plots the z value of each vertex along the selected line to create a profile of the surface at those points. By imagining a direct line of sight from a sitting position on the lakeview bench at 854 feet, you can see that a viewer would be unable to see the slope dropping down to the pond's edge, and also unable to see the ground at the location of the barn, where the surface levels to a flat plane. You can also see that if the barn is tall enough, the viewer would be able to see the upper portions of it.

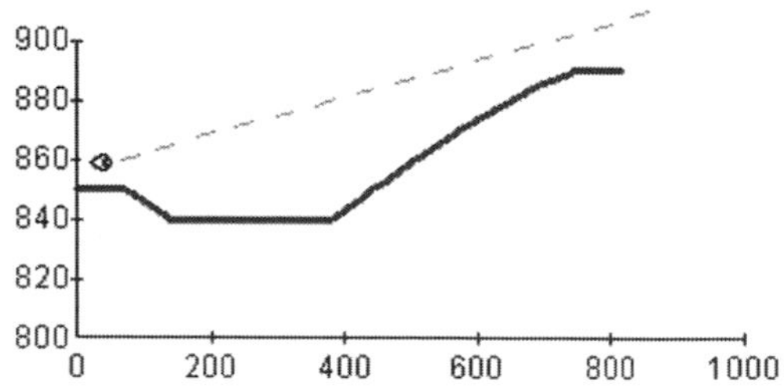

Step 11. Give the Graph a Title

Right-click the title bar of the graph and choose Properties.

In the dialog that appears, click the Appearance tab. Change the title to "View over the Pond" and the subtitle to "to the barn, that is." Or write your own title and subtitle.

Click OK.

Step 12. Change the Name of the Graph

Although the title of the graph was "Profile Graph Title," and you changed it, the name of the graph is also "Profile Graph Title," and you can't change that fact from the Graph Properties dialog. You have to go into the Graph Manager.

From the Tools menu, choose Graphs, and then Manage.

In the Graph Manager, click once on the graph name to highlight it, and then click again to type in the highlighted bar.

Change the name to "Line of Sight" or any other name you prefer.

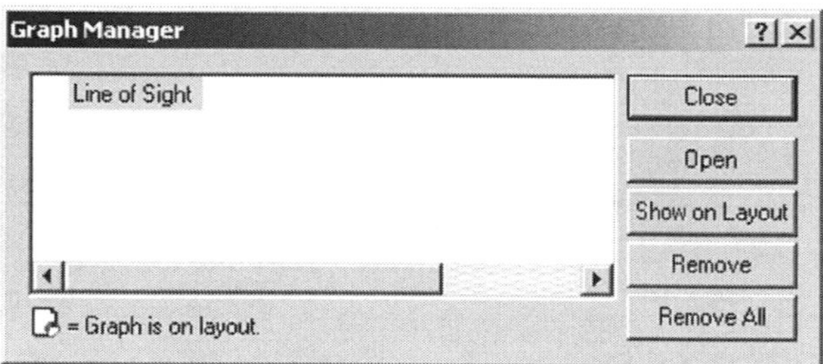

Close the Graph Manager.

Step 13. Close ArcMap

Feel free to experiment with other 3D graphics, lines of sight, and profile graphs. When you're finished, give the ArcMap document a name of your choice and save it in your 3DDATA\Chapter09\MyData folder. Then close ArcMap.

Exercise 9-3

Calculate Surface Area and Volume on a TIN

3D Analyst can calculate the area and volume of a surface above or below an elevation that you specify. In this exercise, you'll calculate the volume of water in the pond on the farm in Carlisle, Kentucky.

Step 1. Start ArcScene and Add Data

Open ArcScene, and click the Add Data button.

Navigate to your 3DDATA\Chapter09\Data folder, and add brushy_creek_tin and lake_tin to the scene.

Click the Add Data button again, and from your Chapter09\MyData folder add 3d_buildings and 3d_lake. (If you don't have these files, use their backup files from the Chapter09\Data folder.)

Step 2. Adjust Vertical Exaggeration

In the Table of Contents, right-click Scene Layers and choose Scene Properties.

On the General tab, change the Vertical Exaggeration to 3. This makes the terrain a little more exciting.

Click OK.

Step 3. Symbolize the Lake and the Buildings

In the Table of Contents, right-click the symbol under 3d_lake. From the color palette choose a light blue shade.

Double-click 3d_buildings in the Table of Contents to bring up its layer properties.

On the Extrusion tab, check the box to extrude the features. Click the Calculator button.

In the Expression Builder, select Ht_Feet from the Fields box. It should appear in the Expression box.

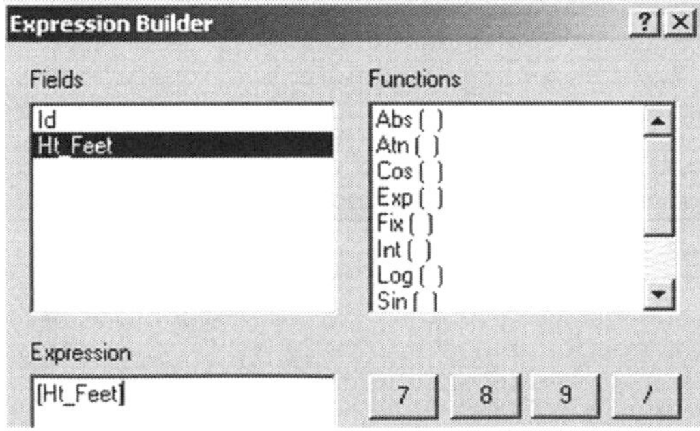

Click OK.

The Extrusion tab should look like this:

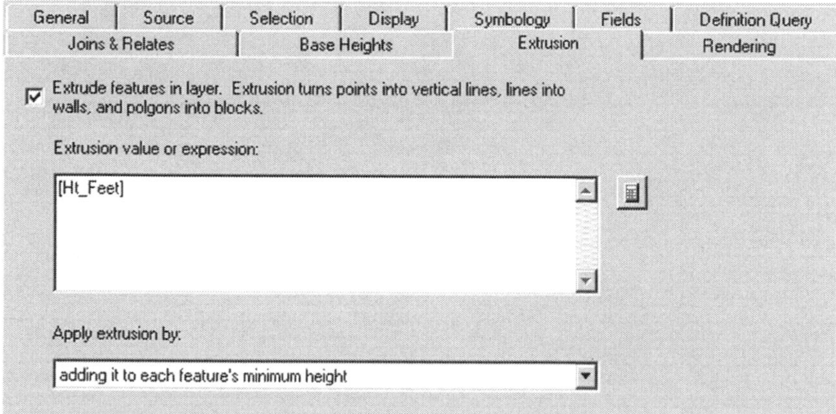

Click OK on the Extrusion tab.

Now the lake is blue, and the buildings are represented by their height in feet. Notice that they appear taller than they did in Exercise 9-1. That's because you added a vertical exaggeration to this scene, and it affects extruded vector features as well the surface.

Step 4. Find the Elevation at the Pond's Edge

In the Table of Contents, right-click 3d_lake and choose Zoom to Layer.

Click the Identify tool. 

Click at various points around the edge of the pond, noting the elevation at each location. It rounds to about 850 feet above sea level.

Close the Identify Results box.

Step 5. Examine Lake_tin

In the Table of Contents, turn off 3d_lake and brushy_creek_tin.

Lake_tin is the surface you used in Exercise 9-1 to digitize 3d_lake.shp. Like brushy_creek_tin, it was originally created from contour lines. The Area and Volume tool calculates statistics for an entire surface, so you need to have a TIN or elevation raster of the precise area that you're interested in.

Step 6. Calculate Area and Volume Statistics

To calculate the amount of water is in the pond, you'll use the elevation around the water's edge (850 feet) as the reference plane, and calculate the volume of lake_tin below that plane.

From the 3D Analyst menu, click Surface Analysis, then Area and Volume.

Change the input surface to 3DDATA\Chapter09\Data\lake_tin.

Change the height of the plane to 850.

Click the button next to "Calculate statistics below plane."

Check the box next to "Save/append statistics to text file." Click the Browse button next to the file path, and navigate to your 3DDATA\Chapter09\MyData folder. Once there, click the Save button.

When you're finished, the Area and Volume Statistics dialog should look like this:

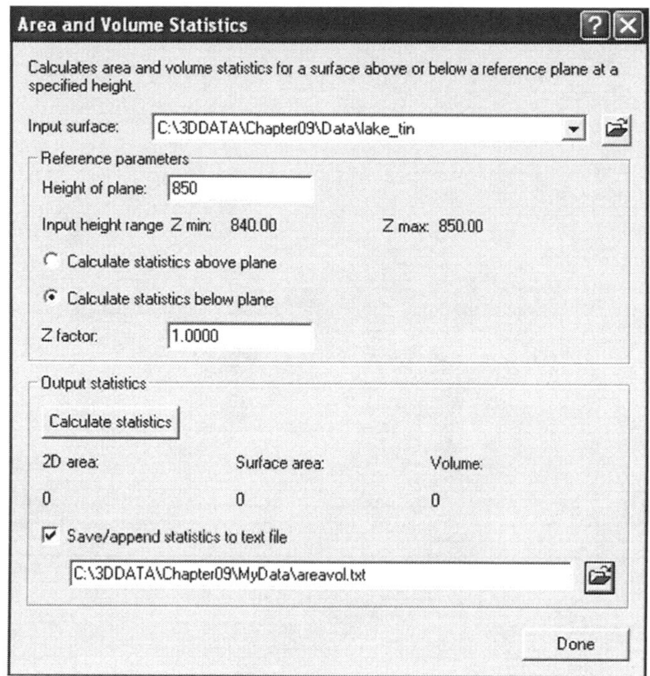

Click the Calculate Statistics button.

The 2D area, surface area, and volume of the pond are reported in the lower panel.

The 2D area corresponds to the flat plane outlined by the upper edge of the lake_tin. It's the same area encompassed by 3d_lake.shp.

The surface area measures the same extent of lake_tin as the 2D area but includes the slopes of the surface. Unless a surface is flat, its surface area will always be greater than its 2D area. A large discrepancy between the measurements indicates that the surface has steep slopes or a lot of variation.

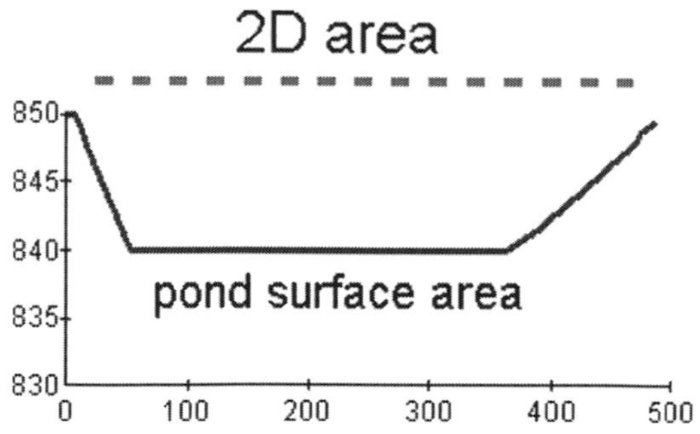

The volume measurement, which is what we're most interested in, is 1,022,780.36 cubic feet.

Click Done.

Step 7. Open the Areavol.txt File

Using a text editor, navigate to your 3DDATA\Chapter09\MyData folder and open areavol.txt. (If you don't have this file in your MyData folder, use its backup from the Chapter09\Data folder.)

Generally, Notepad works well for these text files, but Word does not. If you aren't sure where Notepad is, go to the Start menu on your computer's desktop, then choose Programs, then Accessories. It's usually listed there.

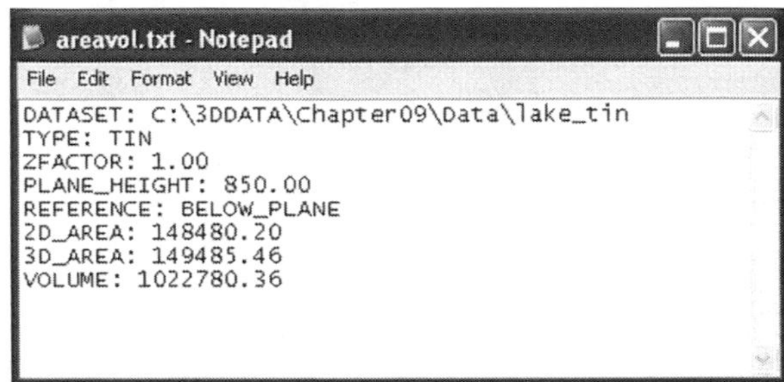

Step 8. Close ArcScene

Give the ArcScene document a name of your choice, and save it in your 3DDATA\Chapter09\MyData folder.

Challenge Exercise

Create Multipatch 3D Features

In Chapter 2, you saw that polygons, whether 2D or 3D, don't drape very well over terrain surfaces because only their vertices contain coordinate and elevation values. We solved that problem in Chapter 7 by integrating polygon features into a TIN, to be displayed as "tag values." There's another option, however, which is to create a multipatch feature class that conforms to a TIN's surface. In some ways, it combines the best of both worlds: the surface conformation of a TIN with the portability of a polygon feature class.

In this exercise, you'll use the TIN of Elk Park that you created in Chapter 7 to interpolate multipatch feature classes of the Elk Park polygon boundary and the Summit polygons.

Step 1. Start ArcScene and Add Data

Open ArcScene, and click the Add Data button.

Navigate to your 3DDATA\Chapter09\Data folder, and add elk_park_bnd.shp, summits.shp, and elk_tin.

Familiarize yourself with the three layers. Turn them on and off in the Table of Contents, and symbolize them with better colors if necessary.

Step 2. Load ArcToolbox

If ArcToolbox isn't visible, click the ArcToolbox button on the Standard Toolbar. 

In ArcToolbox, click the plus sign next to 3D Analyst Tools. Scroll down and expand the TIN Surface toolbox, then double-click "Interpolate Polygon to Multipatch."

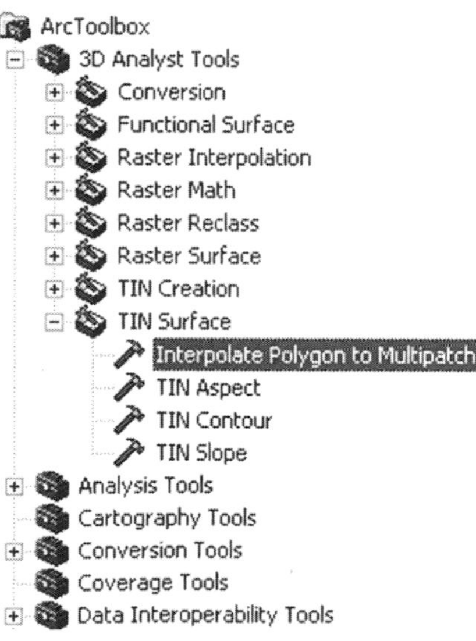

Step 3. Run the Tool

In the dialog, set the Input TIN to elk_tin. The Input Feature Class should be elk_park_bnd. Call the output feature class elk_park_mpatch, and save it in your 3DDATA\Chapter09\MyData folder.

Interpolate Polygon to Multipatch

Input TIN
elk_tin

Input Feature Class
elk_park_bnd

Output Feature Class
C:\3DDATA\Chapter09\MyData\elk_park_mpatch.shp

Maximum Strip Size (optional)
1024

Z Factor (optional)
1

Click OK.

After a while (perhaps a long while) the new multipatch feature class is created. It looks just like elk_park_bnd, but it conforms to the TIN.

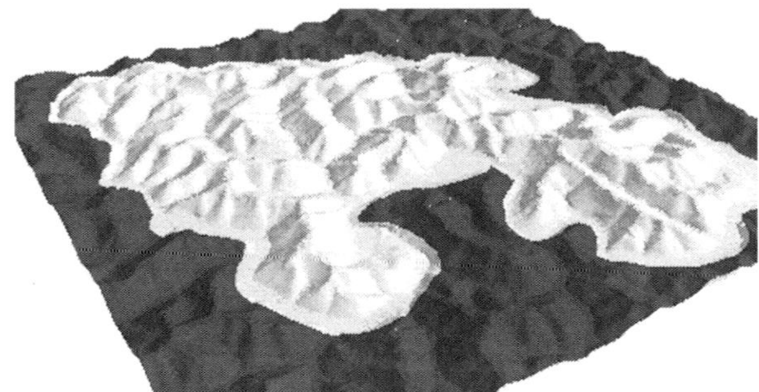

Step 4. Examine the New Layer's Attribute Table and Layer Properties

In the Table of Contents, right-click elk_park_mpatch and open its attribute table. Notice that the Shape field indicates that it has a Multipatch M geometry type.

Close the table. Open the Layer Properties, and click the Source tab. The dialog confirms the geometry type, and also tells you that it's a shapefile.

Click the Base Heights tab. Elk_park_mpatch has z values, obtained from the TIN when you ran the interpolation tool.

Close the Layer Properties dialog after you've looked at the information in some of the other tabs.

Step 5. Look at the Multipatch Shapefile in ArcCatalog

Open ArcCatalog, and navigate to your Chapter09\MyData folder.

Notice that the new multipatch geometry type has its own icon, just as point, polyline, and polygon geometry types have theirs.

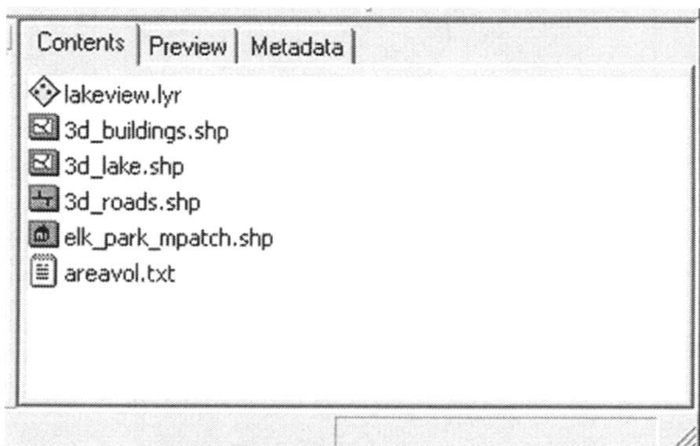

Step 6. Interpolate the Summit Polygons to Multipatch

Follow Steps 2 through 5 again, but use the summits shapefile instead of the elk_park_bnd shapefile as the Input Feature Class in the Interpolate Polygon to Multipatch tool. Call the result summits_mpatch, and save it in your Chapter09\MyData folder.

CHAPTER **10**

SKP to Multipatch to KML: Finalize the Elk Park Project

This chapter is devoted to data conversion between Google Earth, Google SketchUp, and ArcGIS. You'll convert a SketchUp 3D model to a multipatch feature class, and edit it in ArcMap. Then you'll add that multipatch feature class to ArcGlobe and view it as a 3D model in latitude-longitude coordinate space. Finally, you'll export the layers from an ArcMap document to a KML file and view them in Google Earth. If you choose to do the challenge exercise, you'll open a 3D model in SketchUp and export it directly to Google Earth.

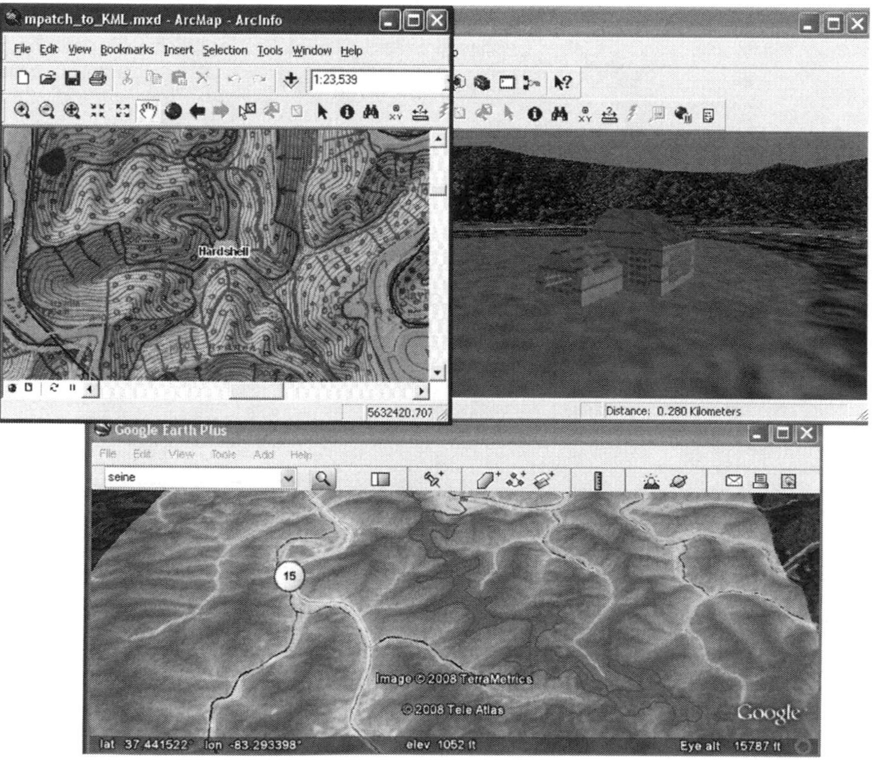

Exercise 10-1

Convert a SketchUp File to a Multipatch Feature Class

Step 1. Open ArcMap, Add Data, and Symbolize It

Start ArcMap and navigate to your 3DDATA\Chapter10\Data folder.

Add o51_landslide.sid, o51_dem, and o51_hshd.

In the Table of Contents, arrange the layers so that o51_landslide.sid is on top, o51_dem is in the middle, and o51_hshd is on the bottom.

Right-click o51_landslide.sid, and choose Properties.

On the Display tab, set the Transparency to 45%. Click OK.

Do the same for o51_dem, and symbolize it with a green color ramp.

Turn the DEM and the Landslide layers on and off. The transparency levels allow the hillshade raster to produce a more dramatic effect, and provide a better sense of the terrain.

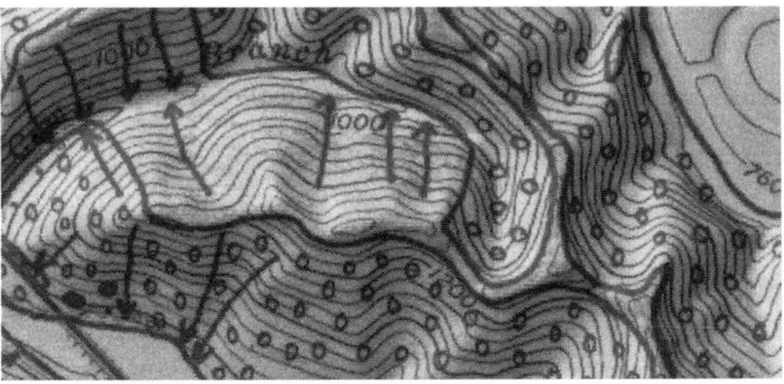

From your 3DDATA\Chapter10\Data folder, add visitor_center.shp, summits.shp, trails.shp, haddix_creeks.shp, haddix_rds.shp, and elk_park_bnd.shp.

You've seen the boundary of Elk Park, the summits, and the creeks and roads before. New to the scene are a polyline shapefile of proposed trails, and a point shapefile that will assist you in creating the Visitor Center multipatch feature.

Symbolize the trails with a red solid line, and choose a line width of 2. The creeks should be bright blue and have a width of 1.5. Roads should have a width of 1.5 and be a solid black. The Elk Park boundary should have a transparent fill and a fairly thick outline—the outline should be a bright color, such as Macaw green. The summits should also be a bright green. You'll be exporting these layers later to KML files to view in Google Earth, so visibility is important.

Step 2. Write Down the X,Y Coordinates of the Visitor Center

In the Table of Contents, right-click visitor_center and choose Zoom to Layer.

Place your cursor on the point feature, and look at the Status Bar at the bottom of ArcMap. (If you don't see the Status Bar, click the View menu and turn it on.)

Write down the coordinates, or if you like just refer to the coordinates in the Status Bar in the graphic below. The x coordinate is listed first, the y coordinate second.

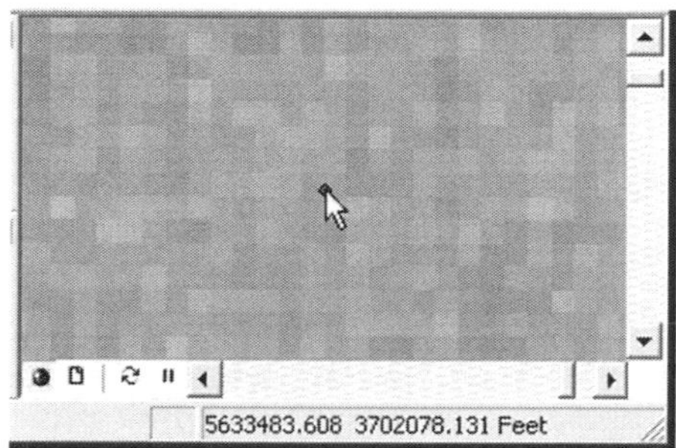

Step 3. Convert a SketchUp 3D Model to a Multipatch Feature Class

In this step, you'll convert visitor_center.skp to a multipatch that can be displayed in ArcScene and ArcGlobe. Visitor_center.skp is a file that I made in about 15 minutes in SketchUp, Google's free 3D modeling program. We don't have time to cover SketchUp in this book, but it is such a popular program and so easy to use that many books and online tutorials have already been written about it. The free version of the program can be downloaded from Google's website, and it comes with simple tutorials that make learning it quite fun. Advanced users have created realistic models of everything from the Eiffel Tower to automobile engines.

If ArcToolbox isn't enabled, click the ArcToolbox button on the Standard toolbar. 

In ArcToolbox, expand 3D Analyst Tools, then Conversion, then From File.

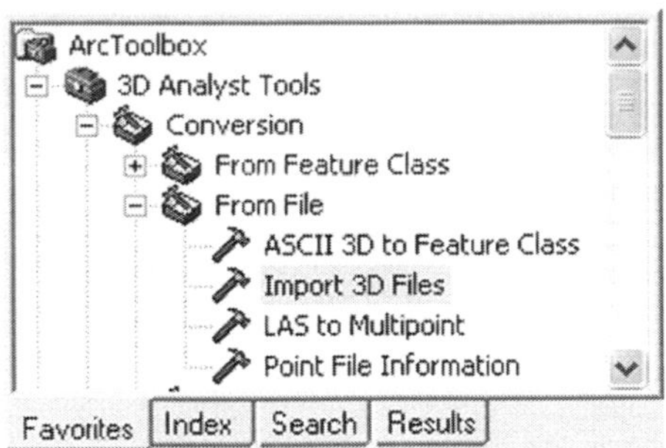

Double-click the Import 3D Files tool to open it.

In the Input Files box, navigate to your 3DDATA\Chapter10\Data folder and add visitor_center.skp.

Usually, you save new files in your MyData folders, but this time you are going to convert the .skp file to a feature class inside a geodatabase feature dataset that's waiting for you in the Chapter10\Data folder. So, in the Output Multipatch Feature Class box, navigate to 3DDATA\Chapter10\Data\ skp_to_mpatch. gdb\dataset.

Name the new feature class visitor_center_mpatch.

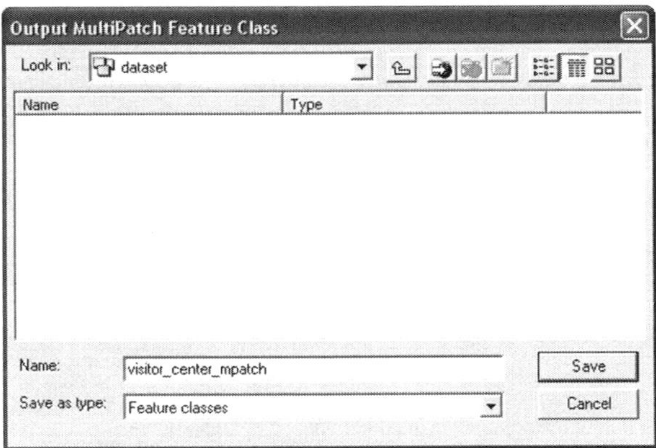

You can leave the Spatial Reference box blank, because the geodatabase dataset already has its coordinate system set to Kentucky State Plane feet.

The Import 3D Files should look like this:

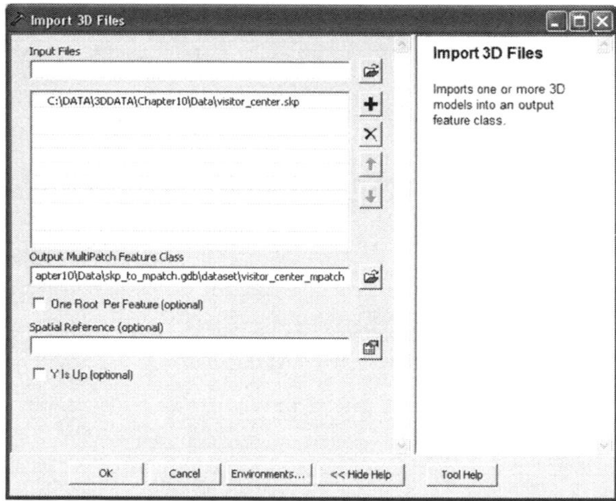

Click OK to run the tool. After a few seconds, the new file is added to ArcMap.

Step 4. Examine the New File

In the Table of Contents, right-click visitor_center_mpatch and select Zoom to Layer.

The new layer is rather disappointing to look at. ArcMap isn't built to display 3D models, so it renders the new multipatch as two flat white rooftops. Not only that, but the layer is off by itself at 0, 0 State Plane feet.

However, visitor_center_mpatch is in the correct coordinate system. In order to scoot it over to its proper location in Elk Park, you need to edit it in ArcMap.

Step 5. Edit the New Multipatch Feature Class

From the View menu, choose Toolbars, and turn on the Editor toolbar.

From the Editor toolbar, choose Start Editing.

When the dialog comes up asking which folder or database you'd like to edit features in, choose 3DDATA\Chapter10\Data\skp_to_mpatch.gdb.

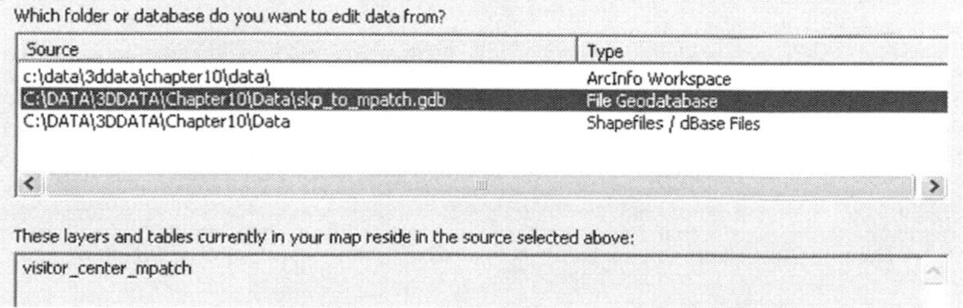

Click OK. (If you get a coordinate system warning, click Start Editing. Some of these datasets are in KY State Plane and some are in Lambert Conformal Conic, but they reside in the same coordinate space.)

In the Editor toolbar, set the Target to visitor_center_mpatch, and the Task to "Modify Feature."

In the Editor toolbar, click the Edit tool.

You should still be zoomed to visitor_center_mpatch.

Select the feature by clicking on it or by dragging a box around it.

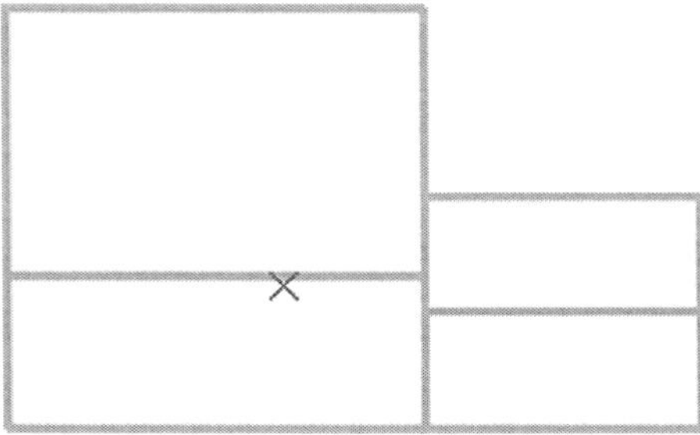

From the Editor dropdown menu, choose Move.

Type in the x,y values that you wrote down (or use the ones in the graphic).

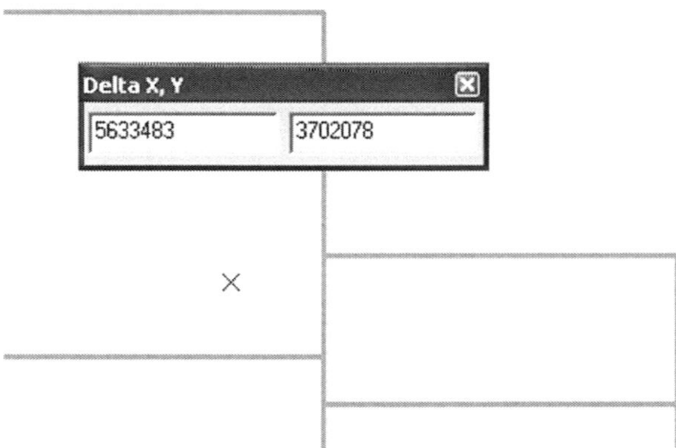

Press the Enter key to give the layer its new coordinates.

The layer disappears.

From the Editor menu, choose Stop Editing. Say Yes to Save Edits.

It may take a few minutes for ArcMap to process the new coordinates, so be patient.

When the hourglass is gone, zoom to the original visitor center point shapefile. You should see the new multipatch feature in the immediate vicinity.

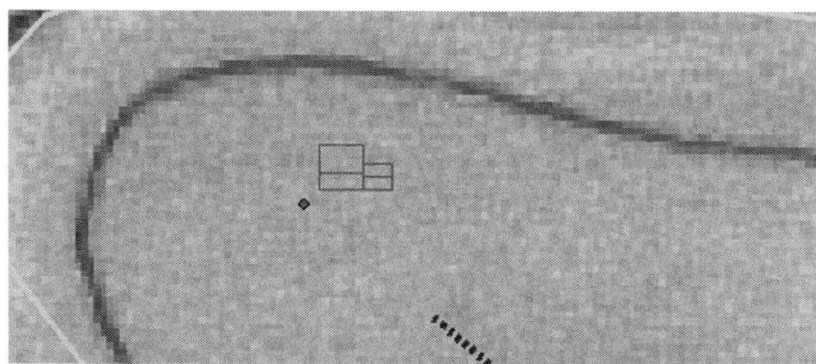

Note that there is another way to convert a SketchUp (.skp) file to a multipatch feature class. That method involves symbolizing a point feature (such as visitor_center.shp) with a 3D marker symbol and choosing the .skp file as that 3D symbol. Then you save the symbolized point feature as a .lyr file and use the ArcToolbox tool "Layer 3D to Feature Class" to convert the .lyr file to a new multipatch feature class. The advantage of that method is that the x,y coordinates are already in place in the original point feature. The advantage of the method that we used in this exercise, however, is that the .skp file is converted directly to a multipatch feature, with no intermediate format change.

Step 6. Save the ArcMap Document

From the File menu in ArcMap, click Save. Name the map "mpatch_to_KML.mxd" and save it in your 3DDATA\Chapter10\MyData folder. You'll be using it again in Exercise 10-3.

Exercise 10-2

View the Multipatch Feature Class in ArcGlobe

Step 1. Start ArcGlobe and Add Data

Open ArcGlobe. In the Table of Contents, right-click Globe Layers, choose Add Data, then Add Draped Data.

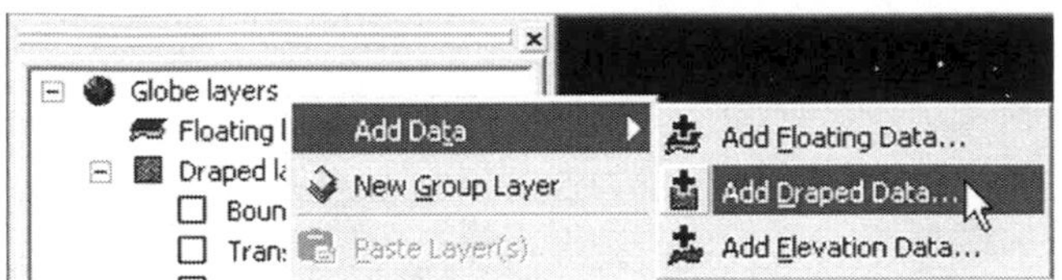

Navigate to your 3DDATA\Chapter10\Data folder.

Add elk_park_bnd, summits, and trails.

When the Coordinate Systems warning comes up, click the Transformations button.

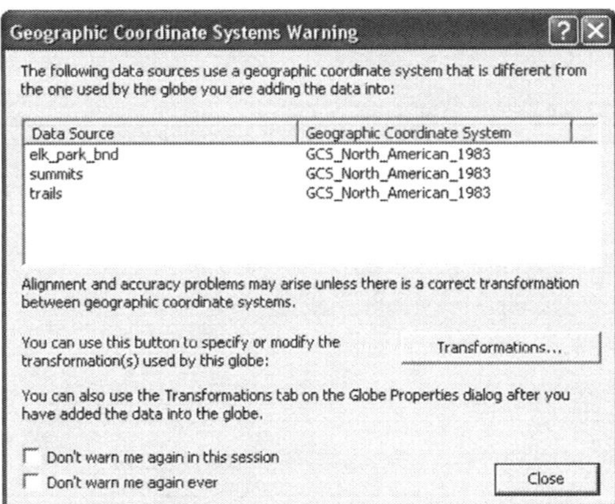

In the next dialog, choose "NAD_1983_to_WGS_1984_5" for the transformation method.

Click OK.

Close the Coordinate Systems warning dialog, and the layers are added to the globe.

Step 2. Symbolize the Layers

In the Table of Contents, drag the Imagery layer so that it is below the three layers that you just added.

Right-click summits, and choose Zoom to Layer.

Symbolize the summits as you like, so that they stand out somewhat from the Imagery layer. Symbolize elk_park_bnd with a transparent fill and a thick outline, again in a color that stands out from the global background. Finally, symbolize the trails with a white line and a width of 2.

Step 3. Add the Visitor Center Multipatch Feature

In the Table of Contents, right-click Globe Layers, choose Add Data, then Add Draped Data.

Navigate to your 3DDATA\Chapter10\Data folder.

Open the skp_to_mpatch geodatabase, then open the dataset, and add visitor_center_mpatch to the globe.

Use the same geographic coordinate transformations settings that you just used with summits, trails, and elk_park_bnd.

Once the multipatch is added, drag it up in the Table of Contents so that it sits above the Imagery layer.

Zoom in so that you can see all of the elements of the scene, and take a look around. Zoom to Layer may not work well with the visitor center, because ArcGIS products seem to have a hard time calculating the bounding box around one coordinate pair. Use the Pan, Zoom, and Zoom to Target tools to zero in on the visitor center.

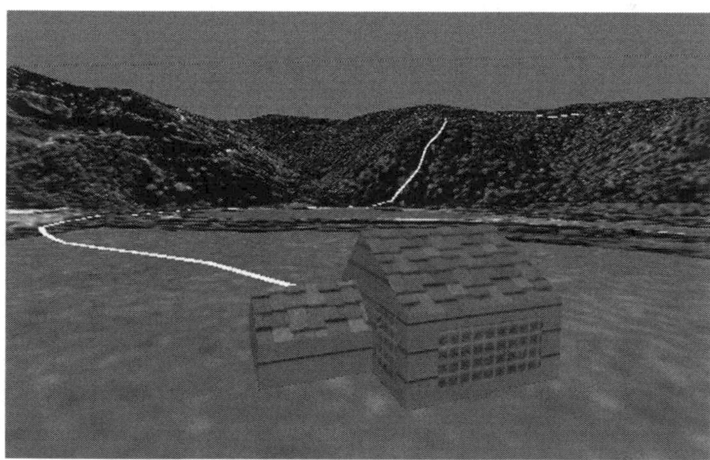

Challenge Step: Add More Data

As this is being written, the Imagery data for ArcGlobe is actually more up to date and detailed for this part of the world than Google Earth's imagery is. If you like, navigate to your 3DDATA\Chapter10\Data folder, and add haddix_rds and haddix_creeks. Experiment with labeling the features, and use the detailed imagery to get an idea of how viable it would be to carve out these proposed trails, with regard to roads, creeks, and general difficulty of terrain.

When you've experimented as much as you like, save the ArcGlobe document with a name of your choice in your Chapter10\MyData folder.

Exercise 10-3

Export Layers from ArcMap to KML, and View Them in Google Earth

Step 1. Open an ArcMap Document

Navigate to your 3DDATA\Chapter10\MyData folder, and open the map document that you made in Exercise 10-1: mpatch_to_KML.mxd.

In the Table of Contents, right-click visitor_center_mpatch, and click Remove. Also remove the visitor_center shapefile.

Make sure that the remaining layers—trails, haddix_creeks, haddix_rds, summits, elk_park_bnd, o51_landslide.sid, o51_dem, and o51_hshd—are turned on in the Table of Contents and that they have the symbology and transparency settings that you specified in Exercise 10-1.

Step 2. Enable HTML Popup Capabilities for Elk_park_bnd

In the Table of Contents, right-click elk_park_bnd and choose Properties.

Click the HTML Popup tab.

You saw in Chapter 5 how Google Earth gives you the option, when you're creating a feature such as a Placemark, a Path, or a Polygon, to write a blurb in the Description panel when you're editing the Properties of the feature. This blurb provides the content for the HTML balloon that pops up in the main viewer when you double-click the name of the feature in the Layers panel in Google Earth's sidebar.

ArcMap, ArcScene, and ArcGlobe also allow you to set HTML Popup properties for layers. You can use a URL that you supply, a customized XSL template, or the fields in the attribute table to populate the popup.

Check the box next to "Show contents for this layer using the HTML Popup tool."

Click the radio button next to "As a table of the visible fields."

Click the Verify button, to get an idea of what a popup will look like for Elk Park.

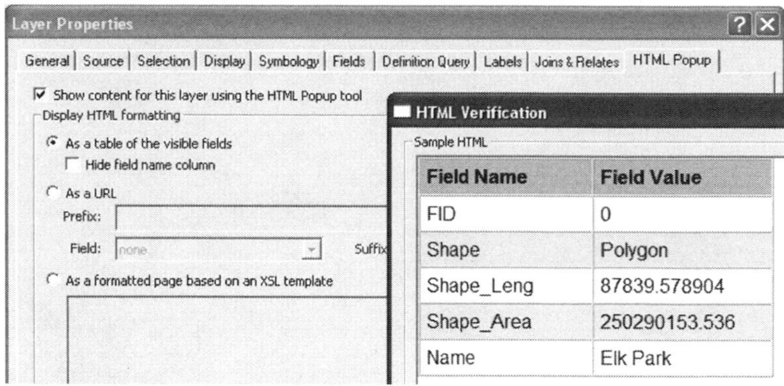

Close the Verify box, and click OK to close the Layer Properties dialog.

Also enable the HTML Popup properties for the Summits shapefile.

Step 3. View the HTML Popups in ArcMap

On the Tools toolbar, click the HTML Popup button. 

Click anywhere inside the Elk Park boundary, or click one of the Summits. The Popup shows the contents of the fields in the attribute table.

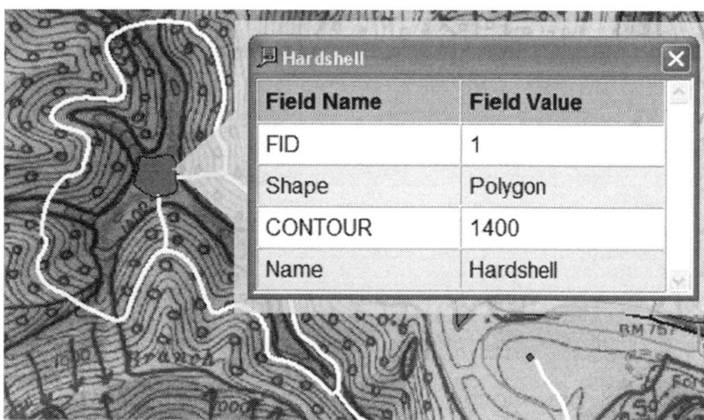

After you've had fun with the HTML Popup tool, select a different tool, such as the Select Elements pointer tool. The HTML Popups can get in the way pretty quickly.

Step 4. Export the Map to a KML File

If ArcToolbox is not visible, click the ArcToolbox button on the Standard toolbar. 

In ArcToolbox, expand 3D Analyst Tools, then Conversion, then To KML.

Open the Map to KML tool by double-clicking on it.

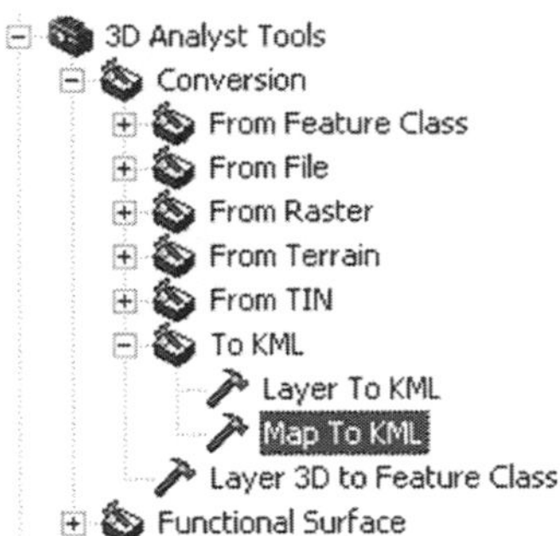

In the Map to KML tool, click the Browse button next to "Map Document" and navigate to your 3DDATA\Chapter10\MyData folder. Add "mpatch_to_KML.mxd."

In the Data Frame box, the entry should be "Layers."

For the Output File, navigate to the Chapter10\MyData folder and call the file "elk_park_map.kmz." (ArcToolbox automatically creates KMZ, not KML files; KMZ is a zipped format that can be opened into its component KML parts in Google Earth.)

Set the Map Output Scale to 100.

Under "Data Content Properties," check the box next to "Convert Vector to Raster."

Under "Extent Properties" click the dropdown arrow and select "Same as layer elk_park_bnd." This is a wonderful feature that lets you control the map extent of the data that is exported.

Leave the Output Image Properties at 1024 and 96.

When you're finished, the dialog should look like this:

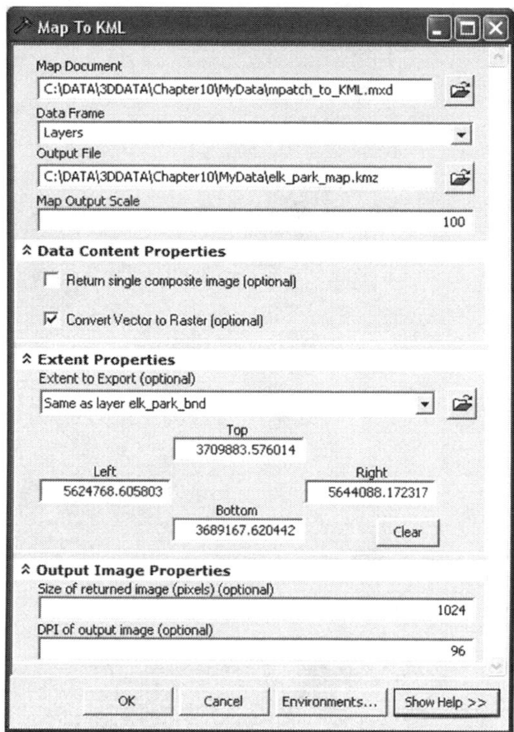

Click OK.

After a few moments, the Map to KML process is complete.

Step 5. Open the New KML File in Google Earth

In Windows Explorer or My Computer, navigate to your 3DDATA\Chapter10\MyData folder.

Double-click elk_park_map.kmz.

Google Earth should start up, and fly to the imaginary Elk Park in Kentucky.

Make sure that in the Google Earth side panel, Roads, Borders, and Terrain are all turned on.

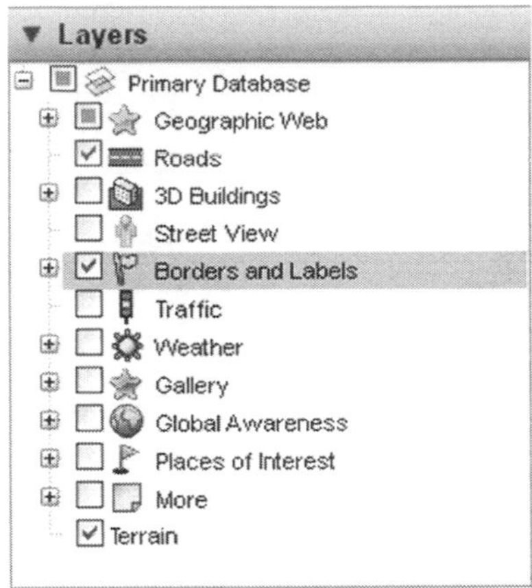

Google Earth puts the new layers in a folder called "Layers" within the Temporary Places folder. Take a look at each of the layers involved in the new KML overlay. Leave o51_hshd turned on, but experiment with the opacity of o51_dem and o51_landslide, turning them on and off to relay different information in the map. (You remember from Chapter 5 that you can change the opacity of a layer by right-clicking it in the "Places" panel and selecting Properties. The Opacity setting is on the Description tab.)

Step 6. Export One Layer to KML

In ArcToolbox, navigate to 3D Analyst Tools > Conversion >To KML.

Open the Layer to KML tool by double-clicking on it.

In the Layer dropdown list, choose trails.

In the Output File box, navigate to your 3DDATA\Chapter10\MyData folder and name the new file "trails_vector.kmz.

Set the Layer Output Scale to 1.

Under Data Content Properties, leave the box unchecked. We are not returning a composite image; we want the trails to be brought over as vector features, not as a rasterized image.

Leave the other settings as they are. The dialog should look like this:

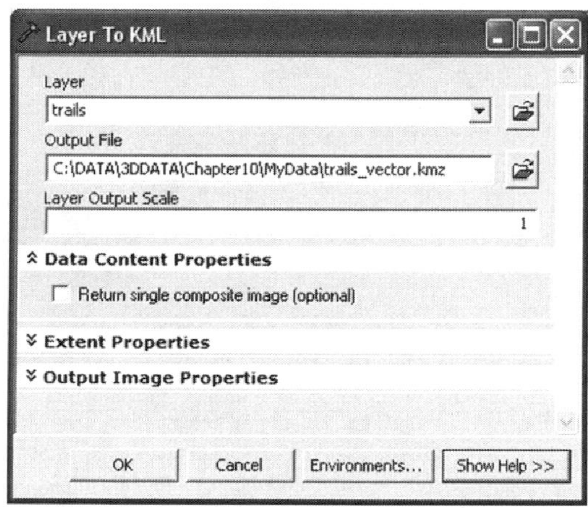

Click OK.

Step 7. Add the New Trails_vector KML File to Google Earth

From the Google Earth File menu, click Open.

Navigate to your Chapter10\MyData folder, and add trails_vector.kmz.

The new trails layer is added to the Places panel in Google Earth. Notice that it isn't called "trails_vector"—the name reflects the layer in ArcMap that was exported, not the name of the .kmz. Also look at the difference between the original "trails" layer that was part of the map export and the new "trails" vector layer that you just added. You can do this by expanding the plus signs next to their names in the Places panel as far as they will go.

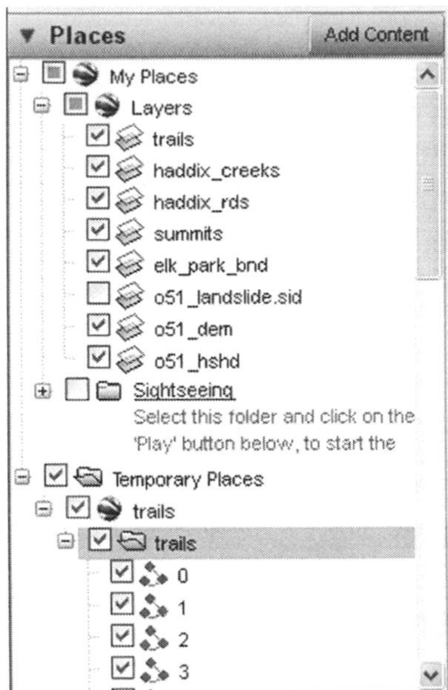

The Map to KML export converted the vector layers—trails, summits, elk_park_bnd, haddix_creeks, and roads—to raster images, and treated them as a "ground overlay" in Google Earth. The Layer to KML export process gave us the option of converting vector to raster, but we left it as vector, and you can see the results in the Places panel. Each feature that had its own record in the attribute table of the trails shapefile is kept intact. They are symbolized in Google Earth as "Path" features.

Turn off the original trails layer by unchecking the box next to it.

Double-click on any one of the individual Path features in the Places panel. Google Earth flies to that particular line segment.

There were quite a few layers involved in the Map to KML export process, so if you have trouble getting one of them to display in Google Earth, you may want to go back to your ArcMap project and use the ArcToolbox "Layer to KML" tool to bring that layer in separately.

You can also experiment with the settings in the "Map to KML" and "Layer to KML" tools, particularly the Data Content Properties setting, which lets you specify whether vector features will be converted to raster in the KML file. Raster is faster and takes up less disk space, but leaving features as vector produces better quality when zooming in. In a close-up view, you can see the difference between the vector trails and the rasterized creeks and summits shapefiles.

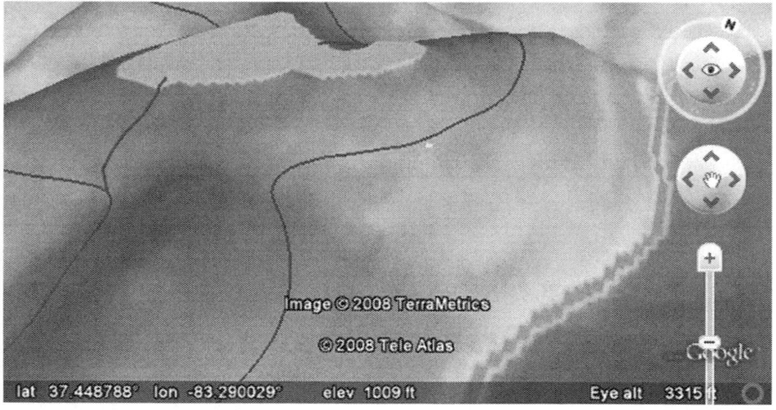

When you've finished checking out all your new layers in Google Earth, save them to the My Places panel upon exiting the program. Exit ArcMap as well.

The next exercise is a quick Challenge Exercise, and in order to go through it you'll need to have SketchUp installed. You'll export the Visitor Center directly from SketchUp into Google Earth.

Challenge Exercise

Export a SketchUp Model to Google Earth

In this chapter, you converted a SketchUp (.skp) file to a multipatch feature inside a geodatabase, using the "Import 3D Files" tool in ArcToolbox. You were then able to view the multipatch in ArcGlobe (and ArcScene).

In Exercise 10-3, you used ArcToolbox to export the "trails" layer to KML. You could have used the same technique to convert the "visitor_center_mpatch" to KML so that it could be viewed in Google Earth, but the results, at least at version 9.3, would be less than stellar. Since Google makes both SketchUp and Google Earth, and since they are designed to work with each other using KML, it makes the most sense to export the visitor center model directly from SketchUp into Google Earth. (In general, it's best to avoid putting your data through unnecessary iterations of non-native software formats.)

If you haven't installed Google SketchUp, the free version can be found at http://sketchup.google.com/

Step 1. Start SketchUp, and Open Visitor_center_bkup.skp

In Windows Explorer or My Computer, navigate to your 3DDATA\Chapter10\Data folder, and double-click "visitor_center_bkup.skp."

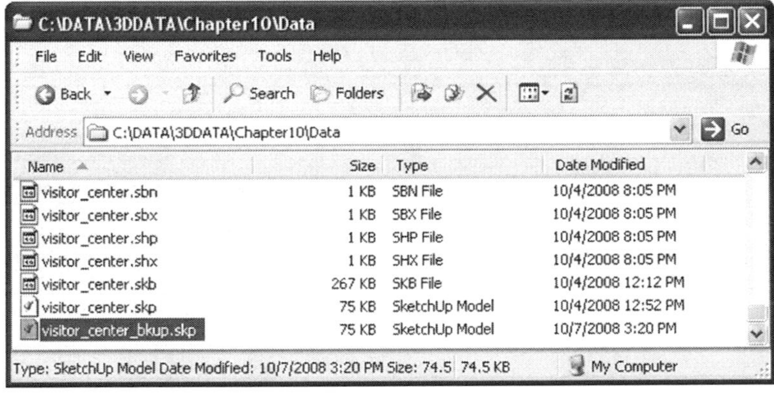

SketchUp starts. (Close the Learning Center startup tip window, if it's in the way.)

You should see the familiar Visitor Center 3D model.

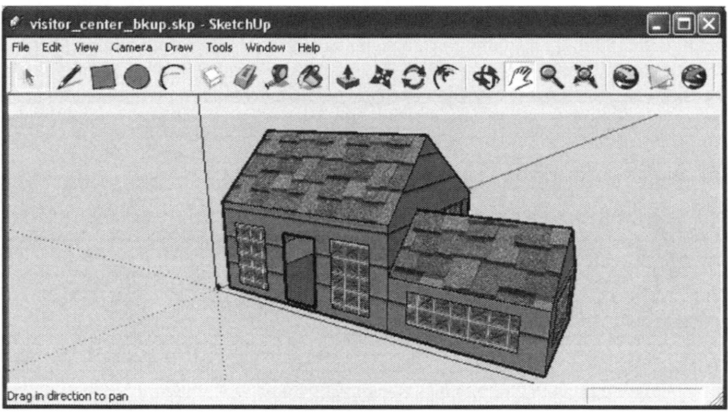

Step 2. Look at the Model with the Orbit Tool

Click the Orbit tool in the toolbar.

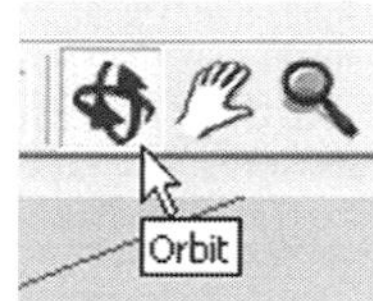

Hold down the left mouse button and move the model around. Use the scroll wheel to zoom in or out. You can view the model from any angle.

Step 3. Open Google Earth, and Fly to Elk Park

Open Google Earth.

In the My Places panel, double-click "Layers" to fly to the vicinity of Elk Park. (If you didn't save "Layers" in My Places in the last exercise, then from the File menu in Google Earth select Open, then navigate to the 3DDATA\Chapter10\MyData folder and add elk_park_map.kmz.)

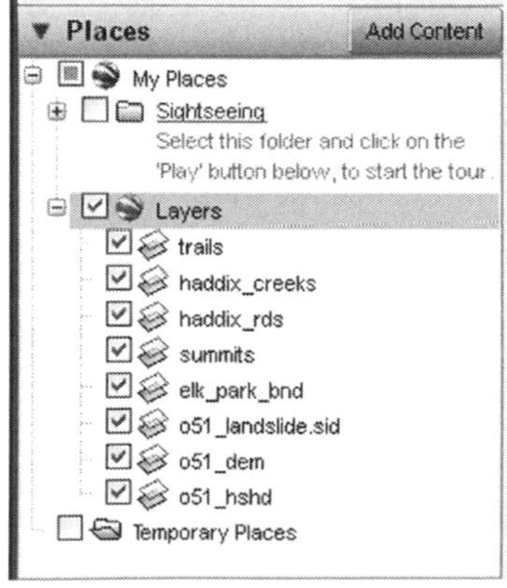

Google Earth zooms in to Elk Park. Once there, zoom in further to the clearing where the Visitor Center should go.

If you'd like exact decimal degree coordinates to fly to in the Search panel, they're

Latitude 37.464576, Longitude–83.2955561.

Zoom in closely—all the way to the clearing. Make sure that you're looking straight down in Google Earth.

Step 4. Bring Google Earth's Coordinates into SketchUp

Back in SketchUp, use the orbit tool to make sure that you're also looking straight down at the model, like this:

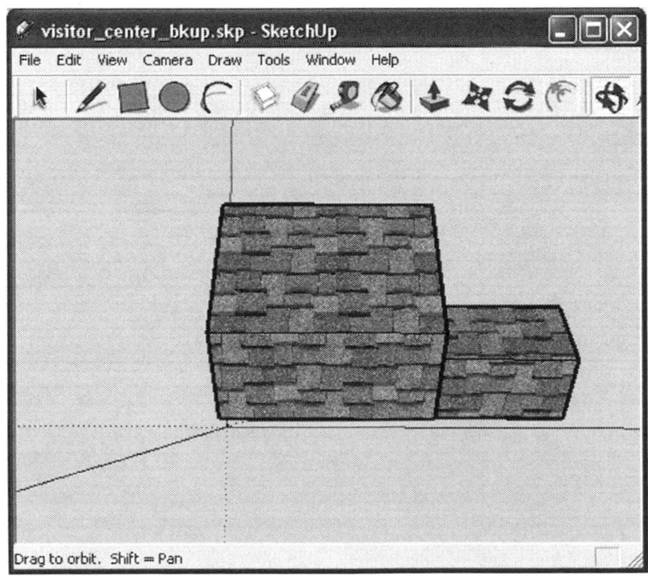

On the SketchUp toolbar, click the Get Current View button.

Google Earth flashes up for a moment, and then SketchUp processes the extent of the current Google Earth view and brings it in to the same space that the model is occupying. It effectively gives the model real-world coordinates.

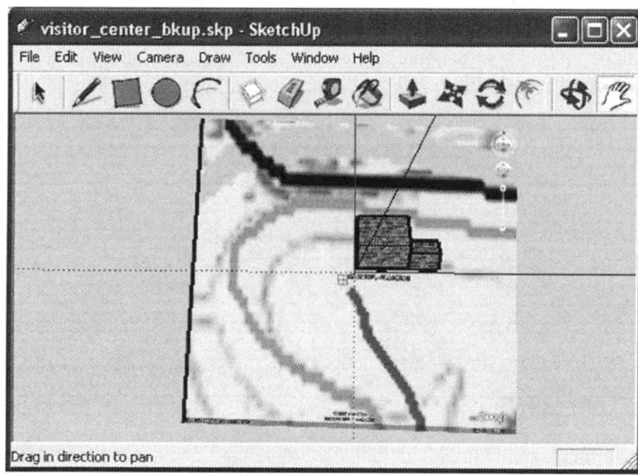

Step 5. Place the Model in Google Earth

On the SketchUp toolbar, click the Place Model button.

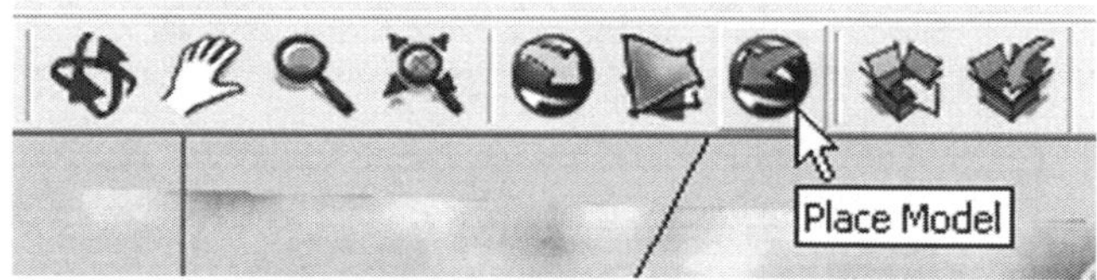

The model is added to Google Earth as an "SU Preview" KML file in the Temporary Places panel.

In the main viewer, take a look at the Visitor Center within the clearing of Elk Park. If the model is too large or too small, you can run through the steps again to get the current view so that the scale will be more realistic.

Experiment more with SketchUp and Google Earth—perhaps converting pertinent layers from ArcMap to KML as vector files instead of rasterized data, so that the creeks, the trails, and the boundary of Elk Park will look good when you fly in close to the 3D model.

When you're finished, you can save the model to My Places upon exiting Google Earth. Don't save changes to the "visitor_center_bkup" SketchUp model, but feel free to save it as a different file name.

About the Tutorial Data

Introduction

This appendix provides you with information about the contents of the archive available at www.wiley.com/college/kennedy. This information is repeated in the introductory passage "Load the Tutorial Data" at the beginning of Chapter 1, Exercise 1-1.

System Requirements

The system requirements to use the contents of this archive are the same requirements for running ArcGIS 3D Analyst. ***You must already have ArcGIS 3D Analyst installed to use this tutorial, as the book does not come with any trial software.*** Most of the 3D Analyst exercises can be done with the 9.1 and 9.2 versions of ArcView, ArcEditor, or ArcInfo; some exercises require 9.3. To do the exercises in Chapters 5 and 10, you will need to install the free version of Google Earth.

Using the Tutorial Data Archive with Windows

Make sure that your computer meets the minimum system requirements listed in this section. If your computer doesn't match up to most of these requirements, you may have a problem using the contents of the archive.

- ❑ PC running Windows 2000, Vista, or XP
- ❑ An Internet connection

Using the Archive

To access the content from the archive follow these steps:

1. Visit the support website for the book to download the tutorial data at www.wiley.com/college/kennedy. Click on the cover for *Introduction to 3D Data: Modeling with ArcGIS 3D Analyst and Google Earth*. On the next page click the link for the Student Companion Site on the right side of the page and follow the links for the 3DDATA.zip archive. Download the zip file to your desktop and extract the archive to the drive of your choice using a program such as WinZip or 7-Zip (freely available at www.7-zip.org). If you have the disk space, I recommend that you copy it directly under your C: drive, so that the full pathname reads "C:\3DDATA."
2. Once the contents of the 3DDATA.zip archive are copied to your hard drive, you can delete the zip file as you won't need it again.

Customer Care

If you have trouble with the archive please call the Wiley Product Technical Support phone number at (800) 762-2974. Outside the United States, call 1(317) 572-3994. You can also contact Wiley Product Technical Support at www.wiley.com/techsupport. John Wiley & Sons will provide technical support only for installation and other general quality control items. For technical support on the applications themselves, consult the program's vendor or author.

To place additional orders or to request information about other Wiley products, please call (877) 762-2974.

Index

A

B

N

O

P

R

S

T

U

V

W

X

Z

Introduction to 3D Data